Mathematical
Surveys
and
Monographs

Volume 120

# Trace Ideals and Their Applications

## Second Edition

Barry Simon

AMERICAN MATHEMATICAL SOCIETY
Providence, Rhode Island

The first edition of *Trace Ideals and Their Applications* was published by Cambridge University Press (1979).

The second edition contains material that originally appeared in *Spectral analysis of rank one perturbations and applications*, Mathematical Quantum Theory. II. Schrödinger Operators (J. Feldman, R. Froese, and L. M. Rosen, eds.), CRM Proc. Lecture Notes, vol. 8, American Mathematical Society, Providence, RI, 1995, pp. 109–149.

2000 *Mathematics Subject Classification.* Primary 47L30; Secondary 47A40, 81U99.

---

For additional information and updates on this book, visit
**www.ams.org/bookpages/surv-120**

---

**Library of Congress Cataloging-in-Publication Data**

Simon, Barry, 1946–
Trace ideals and their applications / Barry Simon.–2nd ed.
p. cm. — (Mathematical surveys and monographs ; v. 120)
Includes bibliographical references and index.
ISBN 0-8218-3581-5 (alk. paper)
1. Operator theory. 2. Hilbert space. 3. Ideals (Algebra). 4. Mathematical physics. I. Title. II. Mathematical surveys and monographs ; no. 120.

QA329.S55 2005
515′.724—dc22　　2005048059

---

AMS softcover ISBN 978-0-8218-4988-0



∞ The paper used in this book is acid-free and falls within the guidelines
established to ensure permanence and durability.
Visit the AMS home page at `http://www.ams.org/`

12 11 10 9 8 7 6 5 4　　28 27 26 25 24 23

# Contents

# Preface to the Second Edition

In 1977, I gave some lectures at the University of Texas which described the general theory of trace ideals initiated by von Neumann and Schatten (*Trace Ideals and Their Applications*, Cambridge University Press, Cambridge, 1979). Because this theory has many different kinds of applications, the lecture notes I produced at the time were widely used, and I got many requests for information on how to obtain it once it fell out of print.

In 1993, I lectured at a summer school in Vancouver on the theory of applications of rank one perturbations of self-adjoint operators (*Spectral analysis of rank one perturbations and applications* in: Mathematical Quantum Theory. II. Schrödinger Operators, CRM Proc. Lecture Notes, 8, pp. 109–149, American Mathematical Society, Providence, RI, 1995). The two topics have much in common. My interest in each arose in my research in several problems at once. And, of course, rank one perturbations are an extreme case of compact perturbations.

Thus, when I started exploring the possibility of reprinting the *Trace Ideals* book as a second edition, it was natural to combine it with my Vancouver lectures. In preparing this new edition, I had to decide first whether to totally rewrite the material, and I chose not to because the basic theory hasn't changed much. Once I made that decision, I felt it made sense to only lightly edit the material, so, for example, references to theorems or equation numbers would be the same. I fixed typos and made a few references to the addendum, especially where a conjecture had been settled. But I followed the original texts closely so much so that the notation in Chapters 1–10 (the original Texas lectures) and Chapters 11–14 (the Vancouver lectures) are, in a few points, slightly different.

I did add a much better index and an addendum describing some developments since the original notes were written.

It is a pleasure to thank many people who helped on this project: R. Bing, J. Dollard, and J. Gilbert for the invitation to give the original lectures in Texas; and J. Feldman, R. Froese, and L. Rosen for the invitation to lecture in Vancouver. V. Jaksic made useful remarks about the addendum. The proofreading was done while I was a Lady Davis Visiting Professor at the Hebrew University of Jerusalem, and I am grateful for the hospitality of the Mathematics Department there provided by H. Farkas and Y. Last. Because of the form of the original notes, it was necessary to TeX them, a task performed admirably by Cherie Galvez. Finally, I acknowledge Martha's love, which makes it all easier.

Barry Simon
Jerusalem, Israel
February 2005

# Preface to the First Edition

Several years ago, I was working simultaneously on three problems: one concerned scattering of a quantum mechanical particle from a very singular repulsive core [**88**], one involved bounds on the number of negative eigenvalues of $-\Delta + \lambda V$ with the correct behavior as $\lambda \to \infty$ [**296**], and the third involved the structure of the two-dimensional Yukawa quantum field theory [**288, 286**]. The physics and the fundamental mathematical structure of these problems are quite different. But it turned out that the technical tools needed to solve the problems were remarkably similar, so much so that at times I couldn't keep straight which one I was thinking about. Since that time, I have had a great respect and use for a subject that might be called "the hard analysis of compact operators in Hilbert space." I discovered that many of the ideas that I grew so fond of had already been developed by Russian mathematicians and mathematical physicists, particularly the group around M. S. Birman (e.g., [**35, 37, 38, 40, 44, 260**]).

In these lectures, I wish to describe the main ideas and illustrate the tools in a group of specific problems. I am a firm believer in the principle that ideas in analysis should be valued largely by their applicability to other parts of mathematics, so I have included lots of applications chosen from my own specialty of mathematical physics, especially quantum theory. However, I have sufficient faith in these tools that I don't doubt that I would have lots of applications if I worked in some other area of analysis. I warn the reader that there is some overlap with pedagogical presentations I have given elsewhere of bits and pieces of this material (Section VI.5, 6 of [**250**]; the appendix to Section IX.4 of [**251**], the second appendix to Section XI.3 in [**253**], Section XIII.17 in [**254**], and my review article on determinants [**300**]) and that virtually nothing I have to say here is not already in the research literature. For beautiful presentations of some of the material from a somewhat different viewpoint I recommend highly the monograph of Goh'berg-Krein [**134**] and Ringrose [**256**]. In particular, much in Chapters 1–3 follows [**134**]. Many of the results of the Birman school are summarized in the lecture notes of Birman and Solomjak [**45**] which have recently been translated.

Like so much of modern analysis, the material to be described has its roots in the famous paper of Fredholm [**115**] (this deep paper is extremely readable and I recommend it to those wishing a pleasurable afternoon). One of the responses to this paper was a flurry of activity from Hilbert and his school which led eventually to the abstraction of what we now call Hilbert space and the Hilbert-Schmidt operators. In modern notation, this latter is the family of operators, $\mathcal{J}_2$, with $\mathrm{Tr}(A^*A) < \infty$. (I should mention that where I use $\mathcal{J}_p$, one often sees $\mathcal{C}_p$, $\mathfrak{C}_p$, or $\mathcal{B}_p$.) For many years, there were many theorems about operators which are products of two or more Hilbert-Schmidt operators (e.g., [**186**]) until von Neumann and Schatten [**280, 279**] formalized the notion of the trace class, $\mathcal{J}_1$. These two ideals are the analogs in a very real sense of $L^2$ and $L^1$: Below, we will develop the theory of the ideals $\mathcal{I}_p$ and $\mathcal{I}_{p,w}$ which are analogous to the $L^p$ and weak-$L^p$ spaces. (See Stein-Weiss [**319**] for a discussion of weak-$L^p$ spaces.) While this analogy is behind much of what we do, we will not systematically develop it — the theory from this point of view has been developed under the name "non-commutative integration" beginning with pioneering work of Segal [**283**]; see also Kunze [**180**], Gross [**139**], and Nelson [**228**] — this abstract theory is done in the more general context where $\mathcal{B}(\mathcal{H})$, the family of all bounded operators on $\mathcal{H}$, is replaced by a von Neumann algebra having a sufficiently regular trace.

Along the way, we will develop not only the trace on a Hilbert space but also the determinant and thereby methods for solving "explicitly" equations of the form $\phi = \psi + K\phi$ for $\phi$ or $\phi = \lambda K\phi$ for $\lambda$. Another kind of problem which becomes tractable concerns the asymptotic distribution of eigenvalues. For suppose that the $n$-th eigenvalue of a positive self-adjoint operator $B$ goes like $n^\alpha$ for $0 < \alpha < 1$. Then $A = (B+1)^{-1}$ will lie in the weak trace ideal $\mathcal{I}_{p,w}$ where $p = \alpha^{-1}$ and this statement captures most of the fact that the $n$-th eigenvalue of $B$ goes like $n^\alpha$. In Chapter 8, we give a hodge-podge of inequalities which may be useful to the reader who wants to apply those methods.

These notes are based on a series of lectures given at the mathematics department of the University of Texas at Austin in April 1977. It is a pleasure to thank Professors R. Bing, J. Dollard, and J. Gilbert for making possible an enjoyable and profitable visit. As usual, I have profited from numerous discussions with various colleagues at Princeton, most especially M. Aizenman, J. Fröhlich, M. Klaus, E. Lieb, and E. Seiler. Finally, I am indebted to Professors A. Pietsch and J. R. Retherford for valuable correspondence concerning the material in Chapter 10, a subject on which I am something of a novice.

Edison, N.J., 1978

Barry Simon

CHAPTER 1

# Preliminaries

We begin with a number of short topics. Since the first two are fairly standard, we will not give all details. The last topic is rather specialized and, in its discussion, we will use results from Chapters 2 and 3.

## 1.1. Absolute Value and Polar Decomposition

Throughout, all our Hilbert spaces will be *complex* and *separable* (are there any others?) and our inner product is linear in the *second* factor and conjugate linear in the first. Recall that a bounded operator, $A$, is called *positive* (written $A \geq 0$) if and only if $(\phi, A\phi) \geq 0$ for all $\phi$; it then follows by polarization that $A^* = A$. If $A - B \geq 0$, we write $A \geq B$ or $B \leq A$. It can be proven that for any $A \geq 0$, there is a unique $B \geq 0$ with $B^2 = A$ (see [**250**, Section VI.4]); we write $B = \sqrt{A}$. For any bounded operator, $A$, the operator $A^*A$ is positive and one defines

$$|A| = \sqrt{A^*A}$$

We warn the reader that

$$|A + B| \leq |A| + |B| \tag{1.1}$$

is *false* in general. For example, if

$$A = \begin{pmatrix} 1 & 1 \\ 1 & 1 \end{pmatrix} \qquad B = \begin{pmatrix} 0 & 0 \\ 0 & -2 \end{pmatrix}$$

then

$$|A + B| = \begin{pmatrix} \sqrt{2} & 0 \\ 0 & \sqrt{2} \end{pmatrix} \qquad |A| + |B| = \begin{pmatrix} 1 & 1 \\ 1 & 3 \end{pmatrix}$$

So $(\phi, |A + B|\phi) > (\phi, (|A| + |B|)\phi)$ for $\phi = \binom{1}{0}$, showing that (1.1) does not hold.

Notice that $\||A|\phi\|^2 = \|A\phi\|^2$. From this, it is easy (see [**250**, Section VI.4]) to construct uniquely an operator $U$ so that

(i) $A = U|A|$
(ii) $\|U\psi\| = \|\psi\|$ for $\psi \in \overline{\operatorname{Ran}|A|} = (\ker A)^\perp$
(iii) $\|U\psi\| = 0$ for $\psi \in (\operatorname{Ran}|A|)^\perp = \ker A$.

Notice that $|A| = U^*A$.

The formula, $A = U|A|$, is called the *polar decomposition* of $A$.

## 1.2. Compact Operators and the Canonical Decomposition

We call a bounded operator $A$ on a Hilbert space *finite rank* if $\dim(\operatorname{Ran} A) < \infty$ (the dimension of $\operatorname{Ran} A$ is called the *rank* of $A$). A bounded operator, $A$, is called *compact* if and only if it is a *norm limit* of finite rank operators. (For a Hilbert space, but not a general Banach space, this is equivalent to the more usual Riesz definition that $A[\{\phi \mid \|\phi\| \leq 1\}]$ has compact closure.) The following two results are standard:

THEOREM 1.1. *Let $A$ be a compact operator on a Hilbert space, $\mathcal{H}$. Then*

(i) *$\sigma(A)$, the spectrum of $A$, is a set with no non-zero limit point.*
(ii) *Every non-zero $\lambda \in \sigma(A)$ is an eigenvalue of finite (geometric) multiplicity.*
(iii) *For any non-zero $\lambda \in \sigma(A)$, there is a finite rank projection $P_\lambda$ so that $AP_\lambda = P_\lambda A$ with $\sigma(A \restriction P_\lambda \mathcal{H}) = \{\lambda\}$, and $\sigma(A \restriction (1 - P_\lambda)\mathcal{H}) = \sigma(A)\backslash\{\lambda\}$. $\dim P_\lambda$ is called the (algebraic) multiplicity of $\lambda$.*

THEOREM 1.2. *Let $A$ be a self-adjoint compact operator on a Hilbert space, $\mathcal{H}$. Then $\mathcal{H}$ has an orthonormal basis of eigenvectors for $A$.*

Theorem 1.2 and (i) and (ii) of Theorem 1.1 are fairly standard and discussed in almost every functional analysis text (cf. [**250**, Section VI.5]). (iii) of Theorem 1.1 is a little more specialized but follows from results in Kato [**164**] or by combining results from [**250**, Section VI.5] and the appendix to [**254**, Section XII.1]. A formula for $P_\lambda$ is

$$P_\lambda = -(2\pi i)^{-1} \int_{|\mu-\lambda|=\varepsilon} (A - \mu)^{-1}\, d\mu$$

where $\varepsilon$ is any sufficiently small positive number.

Let $\mathcal{I}_\infty$ denote the family of compact operators.

THEOREM 1.3. *$\mathcal{I}_\infty$ is a two-sided ideal closed under taking adjoints. In particular, $A \in \mathcal{I}_\infty$ if and only if $|A| \in \mathcal{I}_\infty$.*

PROOF. The finite rank operators are easily seen to be a two-sided ideal closed under taking adjoints, so $\mathcal{I}_\infty$ has the same property. □

THEOREM 1.4. *Let $A$ be a compact operator. Then $A$ has the norm convergent expansion*

$$A = \sum_{n=1}^{N} \mu_n(A)(\phi_n, \cdot\,)\psi_n \tag{1.2}$$

*(where $N$ is a finite non-negative integer or infinity), each $\mu_n(A) > 0$, $\mu_1(A) \geq \mu_2(A) \geq \cdots$, and the $\{\phi_n\}$ and $\{\psi_n\}$ are (not necessarily complete) orthonormal sets. Moreover, the $\mu_n(A)$ are uniquely determined and the $\phi$'s and $\psi$'s are essentially uniquely determined.*

REMARK. (1.2) is shorthand for

$$A\eta = \sum_{n=1}^{N} \mu_n(A)(\phi_n, \eta)\psi_n$$

PROOF. $|A| = U^*A$ is compact and self-adjoint so, by Theorem 1.2, it has a norm convergent expansion

$$|A| = \sum_{n=1}^{N} \mu_n(A)(\phi_n, \cdot\,)\phi_n$$

where the $\mu_n(A)$ are the non-zero eigenvalues of $|A|$ and $\phi_n$ the corresponding eigenvectors. Since $U$ is an isometry on $\operatorname{Ran}|A|$, the $\psi_n = U\phi_n$ are orthonormal, proving (1.2). The uniqueness follows if one notes that if (1.2) holds with the stated properties, then $\mu_n(A)^2$ are the non-zero eigenvalues of $A^*A$, $\{\phi_n\}$ are the eigenvectors of $A^*A$, and $\{\psi_n\}$ are the eigenvectors of $AA^*$. The lack of uniqueness of the $\phi$'s and $\psi$'s comes from the possibility of degenerate eigenvalues of $A^*A$ and $AA^*$. □

The $\mu_n(A)$ are called the *singular values* of $A$. (1.2) is called the *canonical expansion* for $A$.

## 1.3. Inequalities on Singular Values, I

The basis of the inequalities on singular values we discuss now is the elementary equality

$$\mu_n(A) = \mu_n(A^*) \tag{1.3}$$

(which follows from (1.2), or alternatively from the fact that $AA^*$ and $A^*A$ have the same non-zero eigenvalues with the same multiplicities [**87**]) and the following:

THEOREM 1.5.

$$\mu_n(A) = \min_{\substack{\phi_1,\dots,\phi_{n-1} \\ \|\psi\|=1}} \Big[ \max_{\psi \in [\phi_1,\dots,\phi_{n-1}]^\perp} \|A\psi\| \Big]$$

Theorem 1.5 follows from the min-max characterization (see [**254**, Section XIII.1]) of the eigenvalues of $-A^*A$ if we note that $\|A\psi\|^2 = (\psi, A^*A\psi)$ and that $-\mu_n(A)^2$ is the $n$-th eigenvalue of $-A^*A$ counting from the bottom.

THEOREM 1.6. *For any compact $A$ and bounded $B$,*

$$\mu_n(AB) \le \|B\|\mu_n(A) \tag{1.4a}$$

$$\mu_n(BA) \le \|B\|\mu_n(A) \tag{1.4b}$$

PROOF. (1.4a) follows from (1.4b) and (1.3) since (1.3) implies that $\mu_n(AB) = \mu_n(B^*A^*) \le \|B^*\|\mu_n(A^*) = \|B\|\mu_n(A)$. To prove (1.4b), we need only use Theorem 1.5 and $\|BA\psi\| \le \|B\|\,\|A\psi\|$. □

REMARK AND WARNING. $AB$ and $BA$ have the same non-zero eigenvalues but they may not have the same singular values: For example, if $A = \left(\begin{smallmatrix} 0 & 1 \\ 0 & 0 \end{smallmatrix}\right)$, $B = \left(\begin{smallmatrix} 1 & 0 \\ 0 & 0 \end{smallmatrix}\right)$, then $AB = 0$, but $BA = A$, so $\mu_1(AB) = \mu_2(AB) = 0$ but $\mu_1(BA) = 1$, $\mu_2(BA) = 0$.

Theorem 1.6 is the first (take $m = 0$) of a family of inequalities of Fan [**112**] which read for $n, m \ge 0$,

$$\mu_{n+m+1}(AB) \le \mu_{n+1}(A)\mu_{m+1}(B)$$

We will not need these, but the following inequalities of Fan [**112**] will be useful:

THEOREM 1.7 (Fan [**112**]). *Let $A$ and $B$ be compact. Then for $n \ge 0$, $m \ge 0$,*

$$\mu_{n+m+1}(A+B) \le \mu_{n+1}(A) + \mu_{m+1}(B) \tag{1.5}$$

PROOF. Let $Q_n(\phi_1,\dots,\phi_n; A) \equiv \max\{\|A\varphi\| \mid \|\psi\| = 1,\ \psi \in [\phi_1,\dots,\phi_n]^\perp\}$. Since $\|(A+B)\psi\| \le \|A\psi\| + \|B\psi\|$,

$$\begin{aligned} Q_{n+m}(\phi_1,\dots,\phi_{n+m}; A+B) &\le Q_{n+m}(\phi_1,\dots,\phi_{n+m}; A) + Q_{n+m}(\phi_1,\dots,\phi_{n+m}; B) \\ &\le Q_n(\phi_1,\dots,\phi_n; A) + Q_m(\phi_{n+1},\dots,\phi_{n+m}; B) \end{aligned}$$

Minimizing over $\phi_1,\dots,\phi_{n+m}$ and using Theorem 1.5, Fan's inequality (1.5) results. □

## 1.4. Rearrangement Inequalities and All That

DEFINITION. Let $a_n$ be an infinite sequence of numbers with $a_n \to 0$ as $n \to \infty$. $a_n^*$ is the sequence defined by $a_1^* = \max|a_i|$, $a_1^* + a_2^* = \max_{i \neq j}(|a_i| + |a_j|)$, etc. Thus $a_1^* \geq a_2^* \geq \cdots$ and the sets of $a_i^*$ and $|a_i|$ are identical, counting multiplicities.

LEMMA 1.8.

$$\sum_n |a_n b_n| \leq \sum_n a_n^* b_n^*$$

PROOF. By a limiting argument, we may suppose that $a_n, b_n = 0$ for $n \geq N+1$. Without loss, we can renumber the $a$'s so that $|a_1| \geq \cdots \geq |a_N|$. Since

$$\sum_n |a_n b_n| = |a_N| \sum_{n=1}^{N} |b_n| + (|a_{N-1}| - |a_N|) \sum_{n=1}^{N-1} |b_n| + \cdots + (|a_1| - |a_2|)|b_1|$$

and $\sum_{n=1}^k |b_n| \leq \sum_{n=1}^k b_n^*$, we clearly have the required inequality. □

The following result will be central to the treatment in Sections 1.6 and 1.7:

THEOREM 1.9 (essentially due to Markus [**213**]). *Let $a_1 \geq a_2 \geq \cdots \geq a_N \geq 0$ and let $b$ be a point in $\mathbb{C}^N$ with*

$$\sum_{j=1}^{k} b_j^* \leq \sum_{j=1}^{k} a_j \tag{1.6}$$

*for $k = 1, \ldots, N$. Then there exist points $a^{(1)}, \ldots, a^{(m)} \in \mathbb{C}^N$ with $(a^{(\ell)})^* = a$ and $0 \leq \lambda_\ell \leq 1$, $\sum_{\ell=1}^m \lambda_\ell = 1$ so that*

$$b = \sum_{\ell=1}^{m} \lambda_\ell a^{(\ell)}$$

*In particular, if (1.6) holds and $\Phi$ is a function on $([0,\infty))^N$ so that $\phi(c) = \Phi(c_1^*, \ldots, c_n^*)$ is convex on $\mathbb{C}^N$, then*

$$\phi(b) \leq \phi(a)$$

PROOF DUE TO MITJAGIN [**223**]. Let $H$ be the convex hull in $\mathbb{C}^N$ of the points $c$ with $c^* = a$. We want to show that any $b$ obeying (1.6) lies in $H$ (the converse we note is easy). Let $\ell$ be a complex linear function on $\mathbb{C}^N$ and let

$$\beta = \max_{c \in H} \operatorname{Re}(\ell(c))$$

If we prove that $|\ell(b)| \leq \beta$ for each such $\ell$, then $b$ must lie in $H$ by standard hyperplane results (e.g., specialize the Hahn-Banach theorem to $\mathbb{C}^N$). Suppose $\ell(c) = \sum_{n=1}^N \alpha_n c_n$ and let $\ell^*(c) \equiv \sum_{n=1}^N \alpha_n^* c_n$. By Lemma 1.8, $|\ell(b)| \leq \ell^*(b^*)$. Moreover, there exists an $a' \in \mathbb{C}^N$ with $(a')^* = a$ so that $\ell^*(a) = \ell(a')$ and thus $\ell^*(a) \leq \beta$. The proof is thus complete by noting that

$$\begin{aligned}\ell^*(b^*) &= \alpha_N^* \sum_{n=1}^{N} b_n^* + (\alpha_{N-1}^* - \alpha_N^*) \sum_{n=1}^{N-1} b_n^* + \cdots + (\alpha_1^* - \alpha_2^*) b_1^* \\ &\leq \alpha_N^* \sum_{n=1}^{N} a_n + \cdots = \ell^*(a)\end{aligned}$$

□

REMARKS. 1. Results of this genre predate Markus' work considerably. For example, Hardy, Littlewood, and Pólya [**147**] proved that if (1.6) holds with equality for $k = N$, then $b_i^* = \sum \alpha_{ij} a_j$ for a matrix $\alpha$ of positive numbers whose rows and columns sum to 1, and Birkhoff [**33**] has proved that any such $\alpha$ is a convex combination of permutation matrices $\delta_{i\pi(j)}$. Convexity results of the type quoted also go back to Hardy et al. [**147**]. See Addendum A.

2. Theorem 1.9 plays a major role in the Goh'berg-Krein [**134**] presentation of symmetric norms (see Chapter 2) but, strangely enough, they do not use it to obtain the Weyl and Horn inequalities (see Section 1.6 below).

In applications, the $\Phi$'s we will use will either be symmetric norms on $\mathbb{C}^N$ (defined below) or functions of the form $\Phi(x) = \sum_{n=1}^N f(|x_n|)$, where $f$ is convex and monotone increasing on $[0, \infty)$. In the first case, $\Phi$ will be convex by the triangle inequality and homogeneity of norms. In the second, we need only note that for $0 \le \theta \le 1$, $c, d \in \mathbb{C}^N$,

$$\begin{aligned}\phi(\theta c + (1-\theta)d) &= \sum_{n=1}^N f(|\theta c + (1-\theta)d|_n) \\ &\le \sum_{n=1}^N \theta f(|c_n|) + (1-\theta) f(|d_n|) \\ &= \theta\phi(c) + (1-\theta)\phi(d)\end{aligned}$$

(1.6) will arise from two distinct sets of ideas. The first involves:

COROLLARY 1.10. *Let $a_1 \ge a_2 \ge \cdots \ge a_N \ge 0$, $b_1 \ge \cdots \ge b_N \ge 0$. Suppose that*

$$\prod_{j=1}^k b_j \le \prod_{j=1}^k a_j \tag{1.7}$$

*for $k = 1, \dots, N$. Then for any continuous, monotone increasing function $g$ on $[0, \infty)$ with $t \to g(e^t)$ convex, we have that*

$$\sum_{j=1}^k g(b_j) \le \sum_{j=1}^k g(a_j) \tag{1.8}$$

*In particular (taking $g(x) = x$), (1.6) holds.*

PROOF. By a limiting argument, we may suppose that the $a$'s and $b$'s are all non-zero. But then replacing $a_i$ by $\gamma a_i$, $b_i$ by $\gamma b_i$, and $g(x)$ by $g(\gamma^{-1}x)$ for $\gamma$ large, we can suppose all the $a$'s and $b$'s are bigger than 1. Taking $\tilde{a}_j = \log a_j$, $\tilde{b}_j = \log b_j$, the new variables obey (1.6). Letting $f(x) = g(e^x)$ $(x \ge 0)$ and $\Phi(x) = \sum f(x_n)$ as above, (1.8) follows from Theorem 1.9. $\square$

The second involves the following notion:

DEFINITION. A matrix $\{\alpha_{nm}\}_{1 \le n, m \le N}$ is called *doubly substochastic* (dss) if and only if

$$\sum_{n=1}^N |\alpha_{nm}| \le 1 \qquad m = 1, \dots, N$$

and

$$\sum_{m=1}^{N} |\alpha_{nm}| \leq 1 \qquad n = 1, \dots, N$$

The relevance of this concept to operator theory comes partly from:

PROPOSITION 1.11. *If $\{\phi_n\}, \{\psi_n\}$ are orthonormal sets, then $\alpha_{nm} = |(\phi_n, \psi_m)|^2$ is a dss matrix. More generally, if $\{\eta_n\}$, $\{\gamma_n\}$ are also orthonormal sets, then $\beta_{nm} = (\phi_n, \psi_m)(\eta_m, \gamma_n)$ is a dss matrix.*

PROOF. By Bessel's inequality, $\sum_n |\alpha_{nm}| \leq \|\psi_m\|^2 = 1$. The $\beta$ result follows from the $\alpha$ result and the Schwarz inequality

$$\sum_n |\beta_{nm}| \leq \Big(\sum_n |(\phi_n, \psi_m)|^2\Big)^{1/2} \Big(\sum_n |(\eta_m, \gamma_n)|^2\Big)^{1/2} \qquad \square$$

PROPOSITION 1.12. *Let $\alpha$ be a doubly substochastic matrix. Let $c \in \mathbb{C}^N$ with $a = c^*$ and $b_n = \sum_{m=1}^N \alpha_{nm} c_m$. Then $b$ and $a$ obey (1.6).*

PROOF. By relabelling rows and columns of $\alpha$, we can suppose that $|b_1| \geq \cdots \geq |b_N|$ and $|c_1| \geq \cdots \geq |c_N|$. Then

$$\sum_{n=1}^{k} b_n^* \leq \sum_{n=1}^{k} \sum_{m=1}^{N} |\alpha_{nm}|\, |a_m| = \sum_{m=1}^{N} \gamma_n |a_m|$$

where the $\gamma$'s obey $0 \leq \gamma_m \leq 1$, $\sum_{m=1}^N \gamma_m \leq k$. From these properties of the $\gamma$'s and $a_1 \geq \cdots \geq a_N$, one easily sees that $\sum_{m=1}^N \gamma_m |a_m| \leq \sum_{m=1}^k a_m$. $\square$

REMARK. The point of this proposition is the following: $\alpha$ is dss if and only if $\alpha$ is a contraction on $\mathbb{C}^N$ in both the $\ell_1$ and $\ell_\infty$ norm. This proposition and Theorem 1.9 imply that $\alpha$ is then a contraction in all symmetric norms (see Theorem 1.16 below).

## 1.5. Antisymmetric Tensor Products

We review the basic ideas; see [**250**, Sections II.4 and VIII.10] and [**254**, Section XIII.17] for further details. Given Hilbert spaces $\mathcal{H}_1, \dots, \mathcal{H}_n$, we use $\hom(\mathcal{H}_1, \dots, \mathcal{H}_n)$ to denote the maps $\ell : \mathcal{H}_1 \times \cdots \times \mathcal{H}_n \to \mathbb{C}$ which are multilinear. Given $\phi_1 \in \mathcal{H}_1$, $\phi_1 \otimes \cdots \otimes \phi_n$ denotes the multilinear function

$$\phi_1 \otimes \cdots \otimes \phi_n : (\psi_1, \dots, \psi_n) \mapsto \prod_{i=1}^{n} (\phi_i, \psi_i)$$

$\hom_f(\mathcal{H}_1, \dots, \mathcal{H}_n)$ denotes the algebraic span of the $\phi_1 \otimes \cdots \otimes \phi_n$ in $\hom(\mathcal{H}_1, \dots, \mathcal{H}_n)$. $\hom_f$ supports a unique inner product with

$$\langle \ell, \phi_1 \otimes \cdots \otimes \phi_n \rangle = \ell(\phi_1, \dots, \phi_n) \tag{1.9}$$

Using (1.9), the abstract completion of $\hom_f$ can be realized as a subset of $\hom(\mathcal{H}_1, \dots, \mathcal{H}_n)$ denoted $\mathcal{H}_1 \otimes \cdots \otimes \mathcal{H}_n$. Given maps $A_i : \mathcal{H}_i \to \mathcal{H}_i$, there is a map of $\hom(\mathcal{H}_1, \dots, \mathcal{H}_n)$ into itself defined by

$$[(A_1 \otimes \cdots \otimes A_n)(\ell)](\psi_1, \dots, \psi_n) = \ell(A_1^* \psi_1, \dots, A_n^* \phi_n)$$

which takes $\mathcal{H}_1 \otimes \cdots \otimes \mathcal{H}_n$ into itself and obeys

$$(A_1 \otimes \cdots \otimes A_n)(\phi_1 \otimes \cdots \otimes \phi_n) = (A_1\phi_1) \otimes \cdots \otimes (A_n\phi_n)$$

$\otimes_n \mathcal{H}$ is just $\mathcal{H} \otimes \cdots \otimes \mathcal{H}$ ($n$ times). If $\{\phi_i\}$ is an orthonormal basis for $\mathcal{H}$, then $\{\phi_{i_1} \otimes \cdots \otimes \phi_{i_n}\}_{i_1,\dots,i_n}$ is an orthonormal basis for $\otimes_n \mathcal{H}$.

Given $\psi_1, \dots, \psi_n \in \mathcal{H}$, $\psi_1 \wedge \cdots \wedge \psi_n$ is defined by

$$\psi_1 \wedge \cdots \wedge \psi_n = \frac{1}{\sqrt{n!}} \sum_{\pi \in \sigma_n} (-1)^\pi \psi_{\pi(1)} \otimes \cdots \otimes \psi_{\pi(n)}$$

where $\sigma_n$ is the family of all permutations on $\{1, \dots, n\}$ and $(-1)^\pi$ is the sign of the permutation $\pi$. $\Lambda^n \mathcal{H}$ is the Hilbert-span of the $\psi_1 \wedge \cdots \wedge \psi_n$. $\Lambda^0 \mathcal{H} = \mathbb{C}$. Now, a simple calculation shows that

$$(\phi_1 \wedge \cdots \wedge \phi_n, \psi_1 \wedge \cdots \wedge \psi_n) = \det((\phi_i, \psi_j)_{1 \le i, j \le n}) \tag{1.10}$$

so that if $\{\phi_i\}$ is an orthonormal basis for $\mathcal{H}$, then $\{\phi_{i_1} \wedge \cdots \wedge \phi_{i_n}\}_{i_1 < \cdots < i_n}$ is an orthonormal basis for $\Lambda^n \mathcal{H}$. $A \otimes \cdots \otimes A$ ($n$ times) leaves $\Lambda^n \mathcal{H}$ invariant and is denoted $\Lambda^n(A)$. Note that $\Lambda^n(AB) = \Lambda^n(A)\Lambda^n(B)$.

If $\dim \mathcal{H} = n < \infty$, then $\Lambda^n \mathcal{H} = \mathbb{C}$, and

$$\Lambda^n(A) = \det(A) \tag{1.11}$$

which leads to an easy (and quite standard) proof that $\det(AB) = \det(A)\det(B)$.

This machinery will play a major role in the definition of determinants. It is also the natural framework for the following two results:

THEOREM 1.13 (Horn [**150**]). *For any compact operators $A$ and $B$ and all $N$,*

$$\prod_{n=1}^{N} \mu_n(AB) \le \prod_{n=1}^{N} \mu_n(A)\mu_n(B) \tag{1.12}$$

DEFINITION. Given any compact operator, $A$, we will denote the eigenvalues by $\lambda_1(A), \lambda_2(A), \dots, \lambda_n(A), \dots$ ordered so that $|\lambda_1| \ge |\lambda_2| \ge \cdots$ and so that each eigenvalue is counted up to its algebraic multiplicity.

THEOREM 1.14 (Weyl [**340**]). *For any compact operator $A$ and all $N$,*

$$\prod_{n=1}^{N} |\lambda_n(A)| \le \prod_{n=1}^{N} \mu_n(A) \tag{1.13}$$

PROOF OF THEOREMS 1.13 AND 1.14. Let $C$ be a positive compact self-adjoint operator. Let $\phi_1, \dots, \phi_n, \dots$ be a complete orthonormal set of eigenvectors for $C$. Then $\phi_{i_1} \wedge \cdots \wedge \phi_{i_m}$ ($i_1 < \cdots < i_m$) is a complete set of eigenvectors for $\Lambda^m(C)$ so that

$$\|\Lambda^m(C)\| = \prod_{n=1}^{m} \mu_n(C) \tag{1.14}$$

Since $|\Lambda^m(C)| = \Lambda^m(|C|)$ for any $C$, (1.14) holds for any $C$. (1.12) now follows from $\|\Lambda^m(AB)\| \le \|\Lambda^m(A)\| \, \|\Lambda^m(B)\|$. To prove (1.13), one notices that by using a Jordan normal form in $\operatorname{Ran} P_\lambda$ for $\lambda = \lambda_1, \dots, \lambda_N$, we find a set $\{\eta_n\}_{n=1}^N$ of independent vectors so that $A\eta_n = \lambda_n \eta_n + x_n \eta_{n-1}$ with $x_n = 0$ or 1. Clearly,

$$\Lambda^N(A)(\eta_1 \wedge \cdots \wedge \eta_N) = \left( \prod_{n=1}^{N} \lambda_n \right) \eta_1 \wedge \cdots \wedge \eta_N$$

so $\prod_{n=1}^{N} \lambda_n$ is an eigenvalue of $\Lambda^N(A)$ (since $\eta_1 \wedge \cdots \wedge \eta_N \neq 0$ on account of the independence of the $\eta$'s). (1.13) now follows from (1.14) and the trivial fact that the norm must dominate any eigenvalue. □

## 1.6. Inequalities on Eigenvalues, I

In 1909, Schur [**281**] proved that

$$\sum_{i=1}^{N} |\lambda_i(A)|^2 \leq \sum_{1 \leq i, j \leq N} |a_{ij}|^2$$

for an $N \times N$ matrix. The right side of this inequality is precisely $\sum_{n=1}^{N} |\mu_n(A)|^2$ which we will eventually denote by $\|A\|_2^2$. Not long thereafter, Lalesco [**186**] proved that $\sum_{i=1}^{N} |\lambda_i(BC)| \leq \|B\|_2 \|C\|_2$. For any $A$, we can choose $B = U|A|^{1/2}$, $C = |A|^{1/2}$ and thus obtain

$$\sum_{i=1}^{N} |\lambda_i(A)|^p \leq \sum |\mu_i(A)|^p \tag{1.15}$$

for $p = 1$. (Lalesco dealt with infinite matrices, but, as trace class and singular values had not been invented, he discussed products of Hilbert-Schmidt operators.) Much later, Weyl made a systematic study of this question in terms of relations of eigenvalues and singular values.

THEOREM 1.15 (Weyl [**340**], Horn [**150**]). *Let $\phi$ be a non-negative monotone increasing function on $[0, \infty)$ so that $t \mapsto \phi(e^t)$ is convex. Then for any compact $A$,*

$$\sum \phi(|\lambda_n(A)|) \leq \sum \phi(\mu_n(A))$$

*In particular,* (1.15) *holds. Moreover, for any compact $A$ and $B$,*

$$\sum \phi(\mu_n(AB)) \leq \sum \phi(\mu_n(A)\mu_n(B))$$

PROOF. For the first part, it clearly suffices to prove that

$$\sum_{n=1}^{N} \phi(|\lambda_n(A)|) \leq \sum_{n=1}^{N} \phi(\mu_n(A))$$

for each $N$. This follows from Corollary 1.10 and Theorem 1.14. The second part follows similarly, using Theorem 1.13. □

REMARKS. 1. The first inequality is due to Weyl, the second to Horn.

2. If one just wants (1.15), a proof exists which avoids the use of rearrangement-type ideas [**300, 254**].

## 1.7. Symmetrically Normed Spaces

Let $f$ denote the infinite sequences with only finitely many non-zero elements. Following Schatten [**278**], a norm $\Phi$ on $f$ is called *symmetric* if and only if $\Phi(a) = \Phi(a^*)$ for any $a \in f$, that is, if and only if $\Phi$ is invariant under permutations and the maps $a_n \to e^{i\theta_n} a_n$. A sequence $a_n$ is said to lie in the *maximal space* $s_\Phi$ if and only if $\lim_{n\to\infty} \Phi(a_1, \ldots, a_n, 0, 0, \ldots) \equiv \Phi(a)$ exists and is finite. It is then easy to see that $\Phi$ is a norm on $s_\Phi$. The *minimal space* $s_\Phi^{(0)}$ is defined to be the closure (in $s_\Phi$) of $f$. If $s_\Phi = s_\Phi^{(0)}$, that is, if $f$ is dense in $s_\Phi$, we call $\Phi$ *regular* (called mononormalizing in [**134**]).

EXAMPLE. If $\Phi(a) = (\sum |a_n|^p)^{1/p} = \|a\|_p$, then $s_\Phi$ is just the usual $\ell_p$ space. For $p < \infty$, $\Phi$ is regular. The space $\ell_{p,\infty}$ (weak-$\ell_p$) is those sequences $a$ with

$$\|a\|_{p,w}^* \equiv \sup_n (n^{1/p} a_n^*) < \infty$$

$\|\cdot\|_{p,w}^*$ is not a norm (e.g., if $a = (1, 2^{-1/p}, 0, 0, \dots)$ and $b = (2^{-1/p}, 1, 0, 0, \dots)$, then $\|a\|_{p,w}^* = \|b\|_{p,w}^* = 1$, but $\|a+b\|_{p,w}^* = 1 + 2^{1/p}$). However, for $p \neq 1$, one has the Calderón norm:

$$\|a\|_{p,w} = \sup_n \left( n^{-1+\frac{1}{p}} \sum_{j=1}^n a_j^* \right)$$

which is a symmetric *norm* (as $\sup\{|S|^{1+\frac{1}{p}} \sum_{j \in S} |a_j| \mid S \text{ a finite subset}\}$) obeying

$$\|a\|_{p,w}^* \leq \|a\|_{p,w} \leq \frac{p}{p-1} \|a\|_{p,w}^*$$

$\ell_{p,w}$ is the maximal space associated to $\|\cdot\|_{p,w}$. The minimal space is just those $a \in \ell_{p,w}$ with $\lim_{n\to\infty} n^{1/p} a_n^* = 0$. Thus the Calderón norms are not regular for any $p$.

The properties of minimal and maximal spaces are summarized in:

THEOREM 1.16. *Let $\Phi$ be a symmetric norm. Then*

(a) *If $a_n \to 0$ as $n \to \infty$, then $\Phi(a) = \Phi(a^*)$.*

(b) *If $a_n, b_n \to 0$ as $n \to \infty$ and if $\sum_{n=1}^N b_n^* \leq \sum_{n=1}^N a_n^*$ (in particular if $b_n^* \leq a_n^*$), then $\Phi(b) \leq \Phi(a)$.*

(c) *If $\Phi(1, 0, 0, \dots) = c$, then*

$$c\|a\|_\infty \leq \Phi(a) \leq c\|a\|_1$$

*for any $a \in s_\Phi$.*

(d) *Both $s_\Phi$ and $s_\Phi^{(0)}$ are Banach spaces.*

(e) *If $\alpha$ is a doubly substochastic matrix and $a \in s_\Phi$ (resp. in $s_\Phi^{(0)}$), then $\alpha a$, defined by $(\alpha a)_n = \sum_m \alpha_{nm} a_m$ (the sum being absolutely convergent by* (c)) *is in $s_\Phi$ (resp. in $s_\Phi^{(0)}$) and $\Phi(\alpha a) \leq \Phi(a)$.*

(f) *If $\Phi$ is inequivalent to $\|\cdot\|_\infty$, then $s_\Phi$ only consists of sequences $a_n$ with $a_n \to 0$ at $\infty$.*

(g) *If $s_\Phi = s_\Psi$, then $\Phi$ and $\Psi$ are equivalent norms.*

PROOF. (a),(b) Part (b) for finite sequences follows from Theorem 1.9. If we prove (a), then (b) extends to all sequences. Using (b) for finite sequences, we see that $\Phi(a_1, \dots, a_n, 0, 0, \dots)$ is monotone in $n$ and that $\Phi(a_1, \dots, a_n, 0, 0 \dots) \leq \Phi(a_1^*, \dots, a_n^*, 0, 0 \dots)$ so that $\lim \Phi(a_1, \dots, a_n, 0, 0, \dots)$ always exists and $\Phi(a) \leq \Phi(a^*)$. Given $n$, find $N$ so that $a_1^*, \dots, a_n^*$ are among $|a_1|, \dots, |a_N|$. Then, by (b) for finite sequences, $\Phi(a_1^*, \dots, a_n^*, 0, 0, \dots) \leq \Phi(a_1, \dots, a_N, 0, 0, \dots)$, proving $\Phi(a^*) \leq \Phi(a)$.

(c) Clearly, $\Phi(0, \dots, a_n, 0, \dots) \leq \Phi(a) \leq \sum_n \Phi(0, \dots, a_n, \dots, 0)$ (use (b) for the first inequality, the triangle inequality for the second).

(d) Since $s_\Phi^{(0)}$ is closed in $s_\Phi$, we need only consider the case of $s_\Phi$. Let $a^{(m)} \in s_\Phi$ be Cauchy. By (c), $a_n^{(m)}$ is Cauchy for each $n$, so there exists $a$ so that $a_n^{(m)} \to a_n$ for each $n$. Notice that

$$\begin{aligned}\Phi(a_1 - a_1^{(m)}, \dots, a_N - a_N^{(m)}, 0, \dots) &= \lim_{n\to\infty} \Phi(a_1^{(n)} - a_1^{(m)}, \dots, a_N^{(n)} - a_N^{(m)}, 0, \dots) \\ &\le \lim_{n\to\infty} \Phi(a^{(n)} - a^{(m)})\end{aligned}$$

where the inequality comes from (b). Taking $N$ to $\infty$, it follows that $a \in s_\Phi$ and $\Phi(a - a^{(m)}) \to 0$.

(e) Let $b = \alpha a$ and let $b^{(N)} = (b_1, \dots, b_N, 0, \dots)$. By the argument in Proposition 1.12, $\sum_{j=1}^k b_j^{(N)} \le \sum_{j=1}^k a_j^*$ whence $\Phi(b^{(N)}) \le \Phi(a^*)$ by (b). All that remains is to show that $\alpha$ takes $s_\Phi^{(0)}$ into $s_\Phi^{(0)}$. It clearly suffices to show that $\alpha a \in s_\Phi^{(0)}$ for $a \in f$. Let $a = (a_1, \dots, a_N, 0, \dots)$. Only the first $N$ columns of $\alpha$ count for $\alpha a$. Given $\varepsilon$, pick $M$ so that $\sum_{1>M} |\alpha_{ij}| \le \varepsilon$ for $j = 1, \dots, N$. Then $\Phi(0, \dots, 0, (\alpha a)_{M+1}, \dots) \le \varepsilon \Phi(a)$ so $\alpha a \in s_\Phi^{(0)}$.

(f) If $a_n \in s_\Phi$ and $a_n \nrightarrow 0$, then (b) and a simple argument show that $(1, 1, \dots) \in s_\Phi$ whence $\|a\|_\infty \Phi(1, 1, 1, \dots) \ge \Phi(a)$. This inequality and (c) show that $\Phi$ is equivalent to $\|\cdot\|_\infty$.

(g) By (c), the identity map from $s_\Phi$ to $s_\Psi$ is closed, so $\Phi$ is equivalent to $\Psi$ by the closed graph theorem. □

The very simple duality theory for the $s_\Phi$'s is due to Schatten [**278**]. For any symmetric norm $\Phi$, we define the conjugate norm $\Phi'$ on $f$ by

$$\Phi'(b) = \sup\Big\{ \Big| \sum a_n b_n \Big| \;\Big|\; a \in f,\ \Phi(a) \le 1 \Big\}$$

Clearly, using Theorem 1.16(b), the sup is taken by an $a$ which has no more non-zero elements than $b$. Moreover, using Lemma 1.8, it is easy to see that for $b, c$ fixed with $c = c^*$,

$$\sup\Big\{ \Big| \sum a_n b_n \Big| \;\Big|\; a^* = c \Big\} = \sum b_n^* c_n$$

so that $\Phi'$ is a symmetric norm.

THEOREM 1.17 (Schatten [**278**]). *Fix a symmetric norm $\Phi$. Then,*

(a) $\sum |a_n b_n| \le \Phi(a)\Phi'(b)$

(b) $(s_\Phi^{(0)})^* = s_{\Phi'}$ *(including norm) in the sense that any continuous linear functional on $s_\Phi^{(0)}$ has the form $a \mapsto \sum a_n b_n$ for some $b \in s_{\Phi'}$.*

(c) $s_\Phi^{(0)}$ *(resp. $s_\Phi$) is reflexive if and only if both $\Phi$ and $\Phi'$ are regular.*

PROOF. (a) By Lemma 1.8 and Theorem 1.16(b),

$$\begin{aligned}\sum_1^N |a_n b_n| &\le \sum_1^N a_n^* b_n \\ &\le \Phi(a_1, \dots, a_N, 0, \dots)\Phi'(b_1, \dots, b_N, 0, \dots) \\ &\le \Phi(a)\Phi'(b)\end{aligned}$$

(b) By (a), every $b$ in $s_{\Phi'}$ defines a linear functional. For each $N$, there is an $a \in f$ so that $\Phi(a) = 1$ and $|\sum a_n b_n| = \Phi'(b_1, \dots, b_N, 0, 0, \dots)$ so that $\Phi'(b)$ is the norm of $b$ as a linear functional. Thus, we need only show that any $\ell \in s_\Phi^{(0)}$ is of the requisite form. Since $(\mathbb{C}^n)^* = \mathbb{C}^n$, we can find a sequence $b_n$ with $\ell(a) = \sum a_n b_n$ for any $a \in f$. Moreover, by definition of $\Phi'$, $\Phi'(b_1, \dots, b_n, 0, \dots) = \sup\{\ell(a) \mid a = (a_1, \dots, a_n, 0, \dots),\ \Phi(a) = 1\} \leq \|\ell\|$ so $b \in s_{\Phi'}$. Since $b$ is in $s_{\Phi'}$ and $\ell = b$ on $f$, $\ell = b$ on $s_\Phi^{(0)}$.

(c) If $\Phi$ and $\Phi'$ are regular, $(s_\Phi^{(0)})^{**} = s_\Phi^{(0)}$ by (b). If $\Phi$ is not regular, $s_\Phi \subset (s_\Phi^{(0)})^{**}$, so $s_\Phi^{(0)}$ is not reflexive. And if $\Phi'$ is not regular, $(s_\Phi^{(0)})^{**} = (s_{\Phi'})^*$ is strictly bigger than $s_\Phi = (s_{\Phi'}^{(0)})^*$ since, by the Hahn-Banach theorem, there is an element of $s_{\Phi'}$ vanishing on $s_{\Phi'}^{(0)}$. Finally, if $s_\Psi^{(0)}$ is *not* reflexive, then $s_\Phi = (s_{\Phi'}^{(0)})^*$ cannot be reflexive. □

*Note.* For additional discussion of sequence spaces, see Köthe [**177**] and Lindenstrauss-Tzafriri [**204**]. For additional discussion of weak-$L^p$ spaces, see Stein-Weiss [**319**].

## 1.8. Inequalities on Singular Values and Eigenvalues, II

We collect here a few additional inequalities which follow from Theorem 1.9 and Proposition 1.12 and their infinite-dimensional consequence, Theorem 1.16(e).

THEOREM 1.18. *Let $A$ be a compact operator and let $\{\eta_m\}_{m=1}^\infty$ be an orthonormal family. Then*

$$\|A\eta_m\|^2 = \sum_n \alpha_{mn}\mu_n(A)^2 \tag{1.16}$$

*for a dss matrix $\alpha$. Thus, for any symmetric norm, $\Phi$, with $\mu_n(A)^2 \in s_\Phi$, $\Phi(\|A\eta_m\|^2) \leq \Phi(\mu_n(A)^2)$ and, in particular, for any $p \geq 2$,*

$$\sum_{m=1}^\infty \|A\eta_m\|^p \leq \sum_{m=1}^\infty (\mu_n(A))^p \tag{1.17}$$

PROOF. By the canonical decomposition (1.2), equation (1.6) holds with

$$\alpha_{mn} = |(\Phi_n, \eta_m)|^2$$

$\alpha$ is dss by Proposition 1.11. The $\Phi$-inequality follows from Theorem 1.16(e) and (1.17) is just the $\frac{1}{2}p$ power of the $\Phi$-inequality for the symmetric norm $\Phi(a) = (\sum |a_n|^{p/2})^{2/p}$. ($p \geq 2$ is necessary for this to be a norm.) □

REMARKS. 1. The related inequality $\Phi(\langle \eta_m, A\eta_m\rangle) \leq \Phi(\mu_n(A))$ is a special case of Proposition 2.6 below.

2. The requirement $p \geq 2$ is critical for (1.17) to hold. Indeed, if $\eta_m$ is complete, we have, using (1.2), that

$$\begin{aligned}\sum_{n=1}^\infty \mu_n(A) &= \sum_{m=1}^\infty \langle \eta_m, |A|\eta_m\rangle \\ &\leq \sum_{m=1}^\infty \||A|\eta_m\| \\ &= \sum_{m=1}^\infty \|A\eta_m\|\end{aligned}$$

Unless the $\eta_m$ are all eigenvectors of $A$, the inequality will be strict. More generally, for $p \leq 2$,

$$\sum_{n=1}^{\infty} |\mu_n(A)|^p \leq \sum_{m=1}^{\infty} \|A\eta_n\|^p$$

(See [**134**, p. 95 and references therein] for this result and for the history of Theorem 1.18; see also McCarthy [**219**].)

3. Note that for $p = 2$ and $\eta_m$, a basis equality holds in (1.17) since, in that case, $\sum_m \alpha_{mn} = 1$.

THEOREM 1.19 (Goh'berg-Krein [**134**]). *Let $A$ be a compact operator and let $\{P_j\}_{j=1}^J$ be some family of mutually orthogonal self-adjoint projections. Let $B = \sum_{j=1}^J P_j A P_j$. Let $\Phi$ be a symmetric norm. Then $\mu_n(B) \in s_\Phi$ whenever $\mu_n(A) \in s_\Phi$ and*

$$\Phi(\mu_n(B)) \leq \Phi(\mu_n(A)) \tag{1.18}$$

PROOF. By Theorem 1.16(e), it suffices to prove that

$$\mu_m(B) = \sum_{m=1}^{\infty} \alpha_{mn}\mu_n(A) \tag{1.19}$$

for a dss matrix $\alpha$. Let (1.2) be the canonical decomposition for $A$ and let $B = \sum_{m=1}^{\infty} \mu_m(B)(\gamma_m, \cdot\,)\eta_m$ be the canonical decomposition for $B$. Then

$$\mu_M(B) = (\eta_m, B\gamma_m) = \sum_{j=1}^{J} (P_j\eta_m, AP_j\gamma_m)$$

so that (1.19) holds with

$$\alpha_{mn} = \sum_{j=1}^{J} (P_j\eta_m, \psi_n)(\phi_n, P_j\gamma_m)$$

This $\alpha$ is dss as in the proof of Proposition 1.11, for example,

$$\begin{aligned}\sum_n |\alpha_{mn}| &\leq \Big(\sum_{j,n} |(P_j\eta_m, \psi_n)|^2\Big)^{1/2} \Big(\sum_{j,n} |(\phi_n, P_j\gamma_m)|^2\Big)^{1/2} \\ &\leq \Big(\sum_j \|P_j\eta_m\|^2\Big)^{1/2} \Big(\sum_j \|P_j\gamma_m\|^2\Big)^{1/2} \leq \|\eta_m\|\,\|\gamma_m\| \qquad \square\end{aligned}$$

REMARK. Notice that $\{\mu_n(B)\}$ is the symmetric rearrangement of $\cup_j\{\mu_n(P_jAP_j)\}$. Thus, a special case of (1.18) is the fact that for $P$ a self-adjoint projection and $Q = 1 - P$,

$$\sum_{n=1}^{\infty} |\mu_n(PAP)|^p + |\mu_n(QAQ)|^p \leq \sum_{n=1}^{\infty} |\mu_n(A)|^p \tag{1.20}$$

for any $1 \leq p < \infty$.

Part (b) of the following result is due to Lidskii [**193**].

THEOREM 1.20. (a) *For any pair of compact operators, $A, B$,*

$$\mu_n(A) - \mu_n(B) = \sum_{m=1}^{\infty} \alpha_{nm}\mu_m(A-B) \tag{1.21}$$

*for a dss matrix $\alpha$.*

(b) *For any pair of finite-dimensional self-adjoint matrices, $A, B$, let $\tilde{\lambda}_n(A)$ denote the eigenvalues listed in decreasing order. Then*

$$\tilde{\lambda}_n(A) - \tilde{\lambda}_n(B) = \sum_{m=1}^{N} \beta_{nm}\tilde{\lambda}_m(A-B)$$

*for a dss matrix $\beta$.*

PROOF. We will need some facts from elementary eigenvalue perturbation theory [**164, 254**]. Let $Q(x)$ be a polynomial in $x$ with coefficients which are compact operators, so that for $x$ real and near zero, $Q(x)$ is positive. Then a given non-zero eigenvalue, $\lambda_n(Q(x))$, is either degenerate at $x = 0$ in such a way that the degeneracy is removed to some extent for $x$ small, or else $\lambda_n(Q(x))$ is analytic near $x = 0$, and in that case,

$$\left.\frac{d\lambda_n(Qx))}{dx}\right|_{x=0} = \left(\phi_n, \frac{dQ}{dx}(0)\phi_n\right)$$

where $\phi_n$ is any normalized vector with $Q(0)\phi_n = \lambda_n(Q(0))\phi_n$. In the finite-dimensional case, the same is true for $\tilde{\lambda}_n$ if $Q(x)$ is only assumed self-adjoint. Now fix $C$ and $D$ and let $Q(x) = (C + xD)^*(C + xD)$. Write $C = U|C|$ and pick $\phi_n$ so that $|C|\phi_n = \mu_n(C)\phi_n$. Then, unless degeneracy is removed,

$$\begin{aligned}\left.\frac{d\mu_n(C+xD)}{dx}\right|_{x=0} &= \tfrac{1}{2}\,\mu_n(C)^{-1}\,\frac{d[\mu_n(C+xD)]^2}{dx} \\ &= \tfrac{1}{2}\,\mu_n(C)^{-1}\,\frac{d}{dx}\,\lambda_n(Q(x)) \\ &= \tfrac{1}{2}\,\mu_n(C)^{-1}(\phi_n, (C^*D + D^*C)\phi_n) \\ &= \operatorname{Re}(U\phi_n, D\phi_n) = \sum_{m=1}^{\infty} a_{nm}\mu_m(D)\end{aligned}$$

where $a_{nm} = \operatorname{Re}[(U\phi_n, \eta_m)(\gamma_m, \phi_n)]$ if $D = \sum \mu_m(D)(\gamma_m, \cdot\,)\eta_m$ is the canonical decomposition for $D$. As in Proposition 1.11, $A$ is a dss matrix. Next, let $A(x) = B + x(A - B)$. Then $\mu_n(A(x))$ is continuous for all $x$ and differentiable for all $n$ except for a countable set. For $x$ not in this exceptional set, we have that

$$\frac{d\mu_n(A(x))}{dx} = \sum_{m=1}^{\infty} a_{nm}(x)\mu_m(A-B)$$

by the above calculation. Thus (1.21) holds with $\alpha_{nm} = \int_0^1 a_{nm}(x)\,dx$. Since each $a$ is dss, so is $\alpha$.

The proof of (b) is similar, except that we use

$$\frac{d\tilde{\lambda}_n(A(x))}{dx} = (\phi_n(x), (A-B)\phi_n(x)) = \sum b_{nm}(x)\tilde{\lambda}_m(A-B)$$

where $b_{nm} = |(\phi_n(x), \eta_m)|^2$ and

$$A - B = \sum_{m=1}^{\infty} \tilde{\lambda}_m(A-B)(\eta_m, \cdot\,)\eta_m$$

is the eigenexpansion for $A - B$. $\square$

REMARK. We have not been explicit about the usual kind of consequences of these dss relations; we note, however, the inequality,

$$\sum_{n=1}^{\infty} |\mu_n(A) - \mu_n(B)|^p \le \sum_{n=1}^{\infty} |\mu_n(A-B)|^p \tag{1.22}$$

### 1.9. Clarkson-McCarthy Inequalities

We want to note two remarkable inequalities expressed in terms of $\|A\|_p = (\sum |\mu_n(A)|^p)^{1/p}$:

THEOREM 1.21. (a) *For* $2 \le p < \infty$,

$$\|A+B\|_p^p + \|A-B\|_p^p \le 2^{p-1}(\|A\|_p^p + \|B\|_p^p) \tag{1.23}$$

(b) *For* $1 < p \le 2$ *and* $p' = p/(p-1)$,

$$\|A+B\|_p^{p'} + \|A-B\|_p^{p'} \le 2(\|A\|_p^p + \|B\|_p^p)^{p'/p} \tag{1.24}$$

These inequalities, which we call *Clarkson-McCarthy inequalities*, will only be used once again in these notes, namely, to mildly improve a convergence theorem in Chapter 2. They are non-commutative analogs, due to McCarthy [**219**], of some celebrated estimates of Clarkson [**74**]. We will only prove (1.23), which is the simpler of the two. We use some technical results from Chapters 2 and 3. Our proof of (1.23) depends on the following result of McCarthy [**219**] of independent interest:

THEOREM 1.22. *Fix* $1 \le p$. *Let* $A$ *and* $B$ *be two non-negative self-adjoint operators with finite* $\|\cdot\|_p$. *Then*

$$2^{1-p}\|A+B\|_p^p \le \|A\|_p^p + \|B\|_p^p \le \|A+B\|_p^p \tag{1.25}$$

PROOF. We begin with two preliminaries. Let $X$ be a positive self-adjoint operator and let $\|\phi\| \le 1$. Then for $p \ge 1$,

$$|(\phi, X\phi)|^p \le (\phi, X^p\phi) \tag{1.26}$$

Since the spectral theorem asserts that $(\phi, X^a\phi) = \int_0^\infty x^a\, d\mu(x)$ for a suitable measure $d\mu$, (1.26) follows from Hölder's inequality

$$\left(\int_0^\infty x\, d\mu(x)\right) \le \left(\int_0^\infty x^p\, d\mu(x)\right)^{1/p} \left(\int_0^\infty d\mu(x)\right)^{(p-1)/p}$$

For positive numbers $a$ and $b$,

$$(a+b)^p \le 2^{p-1}(a^p + b^p) \tag{1.27}$$

(1.27) is equivalent to $(\frac{1}{2}(a+b))^p \le \frac{1}{2}a^p + \frac{1}{2}b^p$, which just expresses the fact that $x^p$ is a convex function.

To prove the leftmost inequality in (1.25), pick $\eta_n$ a unit vector with $(A+B)\eta_n = \mu_n(A+B)\eta_n$. Then

$$2^{1-p}\mu_n(A+B)^p = 2^{1-p}[(\eta_n, A\eta_n) + (\eta_n, B\eta_n)]^p$$

$$\begin{aligned} &\leq (\eta_n, A\eta_n)^p + (\eta_n, B\eta_n)^p \qquad &\text{(by (1.27))} \\ &\leq (\eta_n, A^p\eta_n) + (\eta_n, B^p\eta_n) \qquad &\text{(by (1.26))} \end{aligned}$$

Summing over $n$ and using $\sum_n (\eta_n, C^p\eta_n) = \operatorname{Tr}(C^p) = \|C\|_p^p$, we have the first inequality in (1.25).

For the second, we may as well suppose that $A > 0$, since the inequality then holds for general $A$ by limiting arguments (using result of Chapter 2). Let $C = A^{1/2}(A+B)^{-1/2}$, $D = B^{1/2}(A-B)^{-1/2}$. Then $C^*C + D^*D = 1$ so, in particular, $\|C\| \leq 1$. Moreover, $A = C(A+B)C^*$. Now pick $\phi_n$ orthonormal with $A\phi_n = \mu_n(A)\phi_n$. Then, since $\|C^*\phi_n\| \leq 1$, we have, by (1.26), that

$$\begin{aligned} \mu_n(A)^p &= (\phi_n, A\phi_n)^p \\ &= (C^*\phi_n, (A+B)C^*\phi_n)^p \\ &\leq (C^*\phi_n, (A+B)^p C^*\phi_n) \end{aligned}$$

Therefore, using the cyclicity of the trace,

$$\begin{aligned} \|A\|_p^P + \|B\|_p^p &\leq \operatorname{Tr}(C(A+B)^p C^* + D(A+B)^p D^*) \\ &= \operatorname{Tr}((C^*C + D^*D)(A+B)^p) = \|A+B\|_p^p \end{aligned} \qquad \square$$

*Warning.* (1.25) is trivial in the commutative case, but there are trivial results in the commutative case which do not extend. For example, if $f$ and $g$ are real-valued functions so that $\|f + ig\|_1 = \|f\|_1$, then $\|g\|_1 = 0$. But if

$$A = \begin{pmatrix} 0 & 1 \\ 1 & 0 \end{pmatrix} \qquad B = \begin{pmatrix} 0 & i \\ -i & 0 \end{pmatrix}$$

then $\|A + iB\|_1 = \|A\|_1$.

Proof of (1.23). We have that, since $p/2 \geq 1$,

$$\begin{aligned} &2^{p-1}(\|A\|_p^p + \|B\|_p^p) \\ &\quad = 2^{p/2}[2^{(p/2)-1}(\|A^*A\|_{p/2}^{p/2} + \|B^*B\|_{p/2}^{p/2})] \\ &\quad \geq 2^{p/2}\|A^*A + B^*B)\|_{p/2}^{p/2} \qquad &\text{(by (1.25))} \\ &\quad = \|(A+B)^*(A+B) + (A-B)^*(A-B)\|_{p/2}^{p/2} \\ &\quad \geq \|(A+B)^*(A+B)\|_{p/2}^{p/2} + \|(A-B)^*(A-B)\|_{p/2}^{p/2} \qquad &\text{(by (1.25))} \\ &\quad = \|A+B\|_p^p + \|A-B\|_p^p \quad \square \end{aligned}$$

M. Klaus [**170**] has remarked that there is a simple proof of the Clarkson-McCarthy inequalities which results from mimicking the elegant proof that Boas [**49**] gave of Clarkson's original $L^p$ inequalities. Namely, let $\mathcal{J}_p^q$ consist of pairs of operators $A, B$ in $\mathcal{J}_p$ with norm

$$\|(A,B)\|_{p;q} = [\|A\|_p^q + \|B\|_p^q]^{1/q}$$

Then (1.23) asserts that the map

$$T : (A,B) \to (A+B, A-B)$$

is bounded from $\mathcal{J}_p^p$ to $\mathcal{J}_p^p$ with norm $2^{1-p^{-1}}$ for $2 \leq p \leq \infty$, and (1.24) asserts that $T$ is bounded from $\mathcal{J}_p^{p'}$ to $\mathcal{J}_p^{p'}$ with norm $2^{1/p'}$ for $1 \leq p \leq 2$. The cases $p = 1, 2, \infty$

of these inequalities are easy. The full inequalities then follow from the Calderón-Lions interpolation theory [**65, 205, 206, 251**] and the calculation that in their framework, $[\mathcal{I}_{p_0}^{q_0}, \mathcal{I}_{p_1}^{q_1}]_t = \mathcal{I}_{p_t}^{q_t}$ with $p_t^{-1} = tp_1^{-1} + (1-t)p_0^{-1}$ and $q_t^{-1} = tq_1^{-1} + (1-t)q_0^{-1}$.

In Chapter 2, we will define $\mathcal{I}_p$ to be those compact operators $A$ with $\|A\|_p < \infty$. We will see that $\mathcal{I}_p$ is a Banach space with $\|\cdot\|_p$ as norm. Recall:

DEFINITION. A Banach space $X$ is called *uniformly convex* if and only if for all $0 < \varepsilon < 2$, the *modulus of convexity*,

$$\delta(\varepsilon) \equiv \inf\left\{1 - \left\|\frac{x+y}{2}\right\| \;\middle|\; \|x\| = 1,\ \|y\| = 1,\ \|x-y\| \geq \varepsilon\right\}$$

is strictly positive.

If one notes that equality holds in (1.23)/(1.24) when $A$ and $B$ are self-adjoint operators with $\ker A = (\ker B)^{\perp}$, we see that, by these inequalities,

$$\delta(\varepsilon) = 1 - \left[1 - \left(\frac{1}{2}\,\varepsilon\right)^p\right]^{1/p} \qquad \text{for } \mathcal{I}_p,\ p \geq 2$$

$$= 1 - \left[1 - \left(\frac{1}{2}\,\varepsilon\right)^{p'}\right]^{1/p'} \qquad \text{for } \mathcal{I}_p,\ p \geq 2;\ p' = \frac{p}{(p-1)}$$

In particular,

THEOREM 1.23 (Dixmier [**96**] for $p \geq 2$; McCarthy [**219**] for general $p$). *For $1 < p < \infty$, $\mathcal{I}_p$ is uniformly convex.*

Uniform convexity in general spaces is important for two reasons: First, uniformly convex spaces are reflexive — since one can compute $\mathcal{I}_p^*$ fairly easily (see Chapter 2), this is of minor interest here. Second, we have the following standard result:

THEOREM 1.24. *Let $X$ be a uniformly convex Banach space and suppose that $x_n$ converges to $x$ weakly, that is, $\ell(x_n) \to \ell(x)$ for all $\ell \in X^*$ and that $\|x_n\| \to \|x\|$. Then $\|x_n - x\| \to 0$.*

PROOF. If $\|x\| = 0$, we are done. If not, we can replace $x_n$ by $x_n/\|x_n\|$ so we are reduced to the case $\|x_n\| = \|x\| = 1$. By the uniform convexity, it suffices to prove that $\|(x_n+x)/2\| \to 1$. Clearly, $\|(x_n+x)/2\| \leq 1$. Now, by the Hahn-Banach theorem, we can find $\ell \in X^*$ so that $\|\ell\| = \ell(x) = 1$. Then

$$\left\|\frac{(x_n+x)}{2}\right\| \geq \frac{1}{2}\,|\ell(x_n+x)| \to 1 \qquad \square$$

We discuss the consequences of this theorem for $\mathcal{I}_p$ in the next chapter. See also Addendum H.

CHAPTER 2

# Calkin's Theory of Operator Ideals and Symmetrically Normed Ideals; Convergence Theorems for $\mathfrak{I}_P$

In this chapter, we will begin by classifying all two-sided ideals of operators (not necessarily norm closed) in $\mathcal{L}(\mathcal{H})$ following Calkin [**66**]. We will set up a one-one correspondence between ideals and certain sequence spaces. Among these sequence spaces are the symmetrically normed spaces $s_\Phi$ and $s_\Phi^{(0)}$ discussed in Chapter 1; the corresponding ideals will be denoted $\mathfrak{I}_\Phi$ and $\mathfrak{I}_\Phi^{(0)}$. Taking $\Phi = \ell_p$ or $\ell_{p,w}$, we obtain the trace ideals $\mathfrak{I}_p$ and $\mathfrak{I}_{p,w}$ which play a major role in later developments. We then discuss criteria for integral operators to lie in the *trace class*, $\mathfrak{I}_1$, and the Hilbert-Schmidt class, $\mathfrak{I}_2$. Finally, we discuss some convergence theorems for $\mathfrak{I}_p$.

We begin our presentation of Calkin's theory with two elementary propositions:

PROPOSITION 2.1. *Let $\mathfrak{I}$ be a two-sided ideal in $\mathcal{L}(\mathcal{H})$ containing an operator $A$ which is not compact. Then $\mathfrak{I} = \mathcal{L}(\mathcal{H})$.*

REMARK. We emphasize our standing convention that $\mathcal{H}$ is separable since it is essential for the truth of this proposition.

PROOF. Since $|A| = U^*A$, $\mathfrak{I}$ contains a positive self-adjoint operator which is not compact. Let $P_a = P_{[a,\infty)}(|A|)$ be the spectral projection for $|A|$. If each $P_a$ $(a > 0)$ is finite-dimensional, then $|A| = \text{norm-lim}_{a\downarrow 0}|A|P_a$ is compact, so some $P_a$ is infinite-dimensional. Since $|A|^{-1}P_a$ is bounded for $a > 0$, each $P_a$ is in $\mathfrak{I}$. If $P_a$ is infinite-dimensional, there exists an isometry $V$ from $\mathcal{H}$ onto $\operatorname{Ran} P_a$. Then $V^*P_aV = 1 \in \mathfrak{I}$, so $\mathfrak{I} = \mathcal{L}(\mathcal{H})$. □

Henceforth, we suppose $\mathfrak{I} \neq \mathcal{L}(\mathcal{H})$ so, by Proposition 2.1, it must consist of compact operators.

PROPOSITION 2.2. *Let $\mathfrak{I}$ be a two-sided proper ideal in $\mathcal{L}(\mathcal{H})$. Let $A$ and $B$ be compact operators with $\mu_n(B) \le \mu_n(A)$ for all $n$. If $A \in \mathfrak{I}$, then $B \in \mathfrak{I}$.*

PROOF. Let $A = \sum \mu_n(A)(\phi_n, \cdot\,)\psi_n$, $B = \sum \mu_n(B)(\eta_n, \cdot\,)\gamma_n$ be the canonical decompositions of $A$ and $B$. Since $\phi$, $\psi$, $\eta$, $\gamma$ are orthonormal and $\mu_n(B) \le \mu_n(A)$, there exists a partial isometry $D$ with $D^*\phi_n = \eta_n$ and a contraction $C$ with $C\psi_n = \mu_n(B)\mu_n(A)^{-1}\gamma_n$. Since $B = CAD$, the proposition is proven. □

This proposition immediately implies:

COROLLARY 2.3. *If $\mathfrak{I} \neq 0$, then every finite rank operator lies in $\mathfrak{I}$. In particular, the only norm closed, two-sided ideals are $\{0\}$, $\mathfrak{I}_\infty$, and $\mathcal{L}(\mathcal{H})$.*

Since $\mu_n(A) = \mu_n(A^*)$, Proposition 2.2 implies that

COROLLARY 2.4. *Every two-sided ideal in $\mathcal{L}(\mathcal{H})$ is left invariant by the adjoint operation.*

The most important consequence of Proposition 2.2 is that ideals are completely described by a set of sequences. Rather than associate $\mathcal{I}$ just with its family of possible singular values, we take a definition allowing enough sequences to form a vector space:

DEFINITION. Fix an orthonormal set $\phi_n$ in $\mathcal{H}$. Given an ideal, $\mathcal{I} \neq \mathcal{L}(\mathcal{H})$, let $\mathcal{S}(\mathcal{I}) = \{a = (a_1, a_2, \dots) \mid \sum a_n(\phi_n, \cdot)\phi_n \in \mathcal{I}\}$.

Thus $a \in \mathcal{S}(\mathcal{I})$ if and only if $a^*$ is a possible set of singular values for an operator in $\mathcal{I}$.

DEFINITION. Given a sequence space $s$, $I(s)$ is the family of compact operators, $A$, with $(\mu_1(A), \dots) \in s$.

DEFINITION. A *Calkin space* is a vector space, $s$, of sequences $a_n$ with $a_n \to 0$ at infinity, with the *Calkin property*: $a \in s$ and $b_n^* \leq a_n^*$ implies $b \in s$.

The following sets up a one-one correspondence between Calkin spaces and two-sided ideals:

THEOREM 2.5 (Calkin [**66**]). *If $s$ is a Calkin space, then $I(s)$ is a two-sided ideal of operators and $\mathcal{S}(I(s)) = s$. If $\mathcal{I}$ is a two-sided ideal, then $\mathcal{S}(\mathcal{I})$ is a Calkin space and $I(\mathcal{S}(\mathcal{I})) = \mathcal{I}$.*

PROOF. If $\mathcal{I}$ is a two-sided ideal, $\mathcal{S}(\mathcal{I})$ is a vector space since $\mathcal{I}$ is a vector space and it has the Calkin property by Proposition 2.2. Moreover, by construction, $I(\mathcal{S}(\mathcal{I})) = \mathcal{I}$.

Now let $s$ be a Calkin space. If we can show that $I(s)$ is an ideal, then $\mathcal{S}(I(s)) = s$ is evident. By the inequalities (1.4), if $A \in I(s)$ and $B \in \mathcal{L}(\mathcal{H})$, then $AB$ and $BA \in I(s)$ by the Calkin property. All that remains is to show that if $A, B \in I(s)$, then so is $A + B$. By Fan's inequalities, (1.5), we have that

$$\begin{aligned} \mu_{2n+1}(A+B) &\leq \mu_{n+1}(A) + \mu_{n+1}(B) \\ \mu_{2n}(A+B) &\leq \mu_{2n-1}(A+B) \leq \mu_n(A) + \mu_n(B) \end{aligned}$$

As a result, $\mu_n(A+B) \leq (\alpha_n + \beta_n)$ where

$$\begin{aligned} \alpha &= (\mu_1(A), \mu_1(A), \mu_2(A), \mu_2(A), \dots) \\ \beta &= (\mu_1(B), \mu_1(B), \mu_2(B), \mu_2(B), \dots) \end{aligned}$$

so it suffices that $\alpha, \beta \in s$, by the Calkin property. But by the Calkin property again, $(\mu_1(A), 0, \mu_2(A), 0, \dots)$ and $(0, \mu_1(A), 0, \mu_2(A), \dots)$ are in $s$. □

Calkin's theorem clearly sets up some relation between symmetric norms on sequences and ideals since every $s_\Phi$ ($\Phi$ is not equivalent to $\ell_\infty$) is a Calkin space, as is every $s_\Phi^{(0)}$. We let $\mathcal{I}_\Phi \equiv I(s_\Phi)$ and $\mathcal{I}_\Phi^{(0)} \equiv I(s_\Phi^{(0)})$. (*Exception and warning*: By this criterion, we should denote the compacts by $\mathcal{I}_\infty^{(0)}$, but we will use $\mathcal{I}_\infty$.) For $A \in \mathcal{I}_\Phi$, we let

$$\Phi(A) = \Phi(\mu_1(A), \mu_2(A), \dots) \tag{2.1}$$

PROPOSITION 2.6 (Simon [**296**]). *Let $\mathcal{B}$ denote the set of pairs of orthonormal sets $\{\phi_n\}, \{\psi_n\}$. If $A \in \mathcal{I}_\Phi$, then*

$$\Phi(A) = \sup_{\phi,\psi\in\mathcal{B}} \Phi((\phi_n, A\psi_n)) \tag{2.2}$$

*Conversely, if $A$ is compact and the right side of (2.2) is finite, then $A \in \mathcal{I}_\Phi$. If $\Phi$ is inequivalent to $\ell_\infty$, then "A is compact" may be replaced by "A is bounded" in the last statement.*

PROOF. Let $A$ be compact and let $A = \sum \mu_n(A)(\eta_n, \cdot)\gamma_n$ be its canonical expansion. Choosing $\psi_n = \eta_n$ and $\phi_n = \gamma_n$, we see that $(\phi_n, A\psi_n) = \mu_n(A)$, so if the right side of (2.2) is finite, then $A \in \mathcal{I}_\Phi$ and $\Phi(A) \le \sup\dots$. On the other hand, for any $\phi, \psi \in \mathcal{B}$, we have that $(\phi_n, A\psi_n) = \sum a_{nm}\mu_m(A)$ where $a_{nm} \equiv (\phi_n, \gamma_m)(\eta_m, \psi_n)$ is dss by Proposition 1.11. Thus, by Theorem 1.16(e), $\Phi((\phi_n, A\psi_n)) \le \Phi(A)$, proving (2.2).

Finally, suppose that $\Phi$ is inequivalent to $\ell_\infty$, that $A \in \mathcal{L}(\mathcal{H})$, and that the right side of (2.2) is finite. Thus, by Theorem 1.16(f), $(\phi_n, A\psi_n) \to 0$ as $n \to \infty$ for all $\phi, \psi \in \mathcal{B}$. It follows that $A$ is compact (completing the proof) by the following argument: Let $\psi_n$ be an orthonormal set in $(\ker(A))^\perp$ and let $U^*\psi_n = \phi_n$ where $A = U|A|$. Then $\||A|^{1/2}\psi_n\| \to 0$. Choose $\psi_n$ inductively by requiring

$$\||A|^{1/2}\psi_n\| \ge \tfrac{1}{2} \sup_{\substack{\phi\in[\psi_1,\dots,\psi_{n-1}]\\ \|\phi\|=1}} \||A|^{1/2}\phi\| \equiv \tfrac{1}{2}\beta_n$$

and we conclude that $\beta_n \to 0$. But then

$$\left\| |A|^{1/2} - \sum_{n=1}^{N} (\psi_n, \cdot)|A|^{1/2}\psi_n \right\| \to 0$$

as $N \to \infty$. □

The properties of $\mathcal{I}_\Phi$ and $\Phi(\cdot)$ on $\mathcal{I}_\Phi$ are summarized in:

THEOREM 2.7. (a) *$\Phi(\cdot)$ is a norm on $\mathcal{I}_\Phi$ obeying*

$$\Phi(ABC) \le \|A\|\,\|C\|\Phi(B) \tag{2.3}$$

$$\Phi(B) \ge \|B\|\Phi(1, 0, \dots) \tag{2.4}$$

*for all $B \in \mathcal{I}_\Phi$, $A, C \in \mathcal{L}(\mathcal{H})$.*

(b) *$\mathcal{I}_\Phi$ and $\mathcal{I}_\Phi^{(0)}$ are Banach spaces in the norm $\Phi$ and $\mathcal{I}_\Phi^{(0)}$ is the closure in $\mathcal{I}_\Phi$ of the finite rank operators. For any $A \in \mathcal{I}_\Phi^{(0)}$, the canonical decomposition converges in $\Phi$-norm.*

(c) *Any norm on an operator ideal $\mathcal{I}$ obeying (2.3) agrees with a symmetric norm, $\Phi$, on the finite rank operators. $\mathcal{I} \subset \mathcal{I}_{\tilde{\Phi}}$. If $\mathcal{I}$ is Banach in its norm, then $\mathcal{I}_{\tilde{\Phi}}^{(0)} \subset \mathcal{I}$.*

(d) (Non-commutative Fatou lemma) *If $A_m \in \mathcal{I}_\Phi$, $A_m \to A$ weakly, $A$ is compact, and $\sup_n \Phi(A_n) < \infty$, then $A \in \mathcal{I}_\Phi$ and $\Phi(A) \le \sup_n \Phi(A_n)$. If $\Phi$ is inequivalent to $\ell_\infty$, $A$ need not be assumed to be compact a priori.*

REMARK. (a)–(c) are essentially due to Schatten [**278**] and (d) to Goh'berg-Krein [**134**].

PROOF. (a) Since $\Phi$ is a norm on sequences,

$$\Phi((\phi_n, (A+B)\psi_n) \leq \Phi((\phi_n, A\psi_n)) + \Phi((\phi_n, B\psi_n))$$

so (2.2) implies that $\Phi$ is a norm. (2.3) follows from (1.4) and Theorem 1.16(b), and (2.4) from Theorem 1.16(c).

(b) Let $A_m$ be Cauchy in $\Phi$-norm. By (2.4), $A_m$ is Cauchy in operator norm so $\|A_m - A\| \to 0$ for some compact $A$. Moreover,

$$\begin{aligned}\Phi((\phi_n, (A_m - A)\psi_n)_{n\leq N}) &\leq \lim_{k\to\infty} \Phi((\phi_n, (A_m - A_k)\psi_n)_{n\leq N}) \\ &\leq \lim_{k\to\infty} \Phi((\phi_n, (A_m - A_k)\psi_n)) \qquad \text{(by Theorem 1.16(b))} \\ &\leq \lim_{k\to\infty} \Phi(A_m - A_k) \qquad \text{(by (2.2))}\end{aligned}$$

Thus by (2.2), $\Phi(A_m - A) \leq \lim_{k\to\infty} \Phi(A_m - A_k)$, so $A_m \to A$ in $\Phi$-norm. This proves that $\mathcal{J}_\Phi$ is a Banach space.

Let $A$ be finite rank. Then, by Theorem 1.16(e), $(\phi_n, A\psi_n) \in s_\Phi^{(0)}$ for all $\phi, \psi \in \mathcal{B}$. It follows from (2.2) that if $A$ is in the $\Phi$-closure of the finite rank operators, then $(\phi_n, A\psi_n) \in s_\Phi^{(0)}$ for any $\phi, \psi \in \mathcal{B}$. In particular, using a canonical decomposition, $\{\mu_n(A)\} \in s_\Phi^{(0)}$. Conversely, if $\{\mu_n(A)\} \in s_\Phi^{(0)}$, the canonical decomposition will converge in $\Phi$-norm, so $A$ is in the norm closure of the finite rank operators.

(c) In particular, (2.3) implies that the norm is unitary invariant and thus only dependent on the singular values of $A$. Thus, there is a symmetric norm $\tilde{\Phi}$ on sequences with $\Phi(A) = \tilde{\Phi}(\mu_1(A), \dots, \mu_N(A), 0, \dots)$ for finite rank $A$'s. By (2.3), $\Phi(A) \geq \Phi(\sum_{n=1}^N \mu_n(A)(\phi_n, \cdot)\psi_n) = \tilde{\Phi}(\mu_1, \dots, \mu_N, 0, \dots)$, so $\Phi(A) \geq \tilde{\Phi}(A)$, and any $A$ in $\mathcal{J}$ is in $\mathcal{J}_{\tilde{\Phi}}$.

(d) For any $\phi, \psi \in \mathcal{B}$,

$$\begin{aligned}\Phi((\phi_n, A\psi_n)_{n\leq N}) &= \lim_{m\to\infty} \Phi((\phi_n, A_m\psi_n)_{n\leq N}) \\ &\leq \limsup_{m\to\infty} \Phi((\phi_n, A_m\psi_n)) \qquad \text{(by Theorem 1.16(b))} \\ &\leq \sup_m \Phi(A_m)\end{aligned}$$

The result now follows from the end of Proposition 2.6. □

REMARKS. 1. In the context of (c), it can happen that $\Phi$ and $\tilde{\Phi}$ are not equal; for example, take $\Phi(A) = \|\mu_n\|_{p,w} + \limsup_{n\to\infty} n^{1/p}\mu_n(A)$, but as in Theorem 1.16(g), if $\mathcal{J}$ has two Banach spaces norm obeying (2.3), they must be equivalent.

2. The triangle inequality for $\Phi$ and $\mathcal{J}_\Phi$ has two other rather different proofs which are instructive. One [**278**] first establishes the formula ($\Phi'$ = conjugate norm to $\Phi$),

$$\Phi(A) = \sup\{|\mathrm{Tr}(AB)| \mid \Phi'(B) = 1\}$$

from which $\Phi(A + C) \leq \Phi(A) + \Phi(C)$ is easy. The other [**134**] first establishes that $\|A + B\|_1 \leq \|A\|_1 + \|B\|_1$ for finite rank operators. Given this and $A + B = \sum_n \mu_n(A + B)(\phi_n, \cdot)\psi_n$, let $P_N$ be the projection onto the space spanned by $\psi_1, \dots, \psi_N$. Then $\|P_N(A + B)\|_1 = \sum_{n=1}^N \mu_n(A + B)$ while $\mu_n(P_N A) \leq \mu_n(A)$ for $n = 1, \dots, N$ and $\mu_n(P_N A) = 0$ for $n > N$. Thus

$$\sum_{n=1}^N \mu_n(A + B) \leq \sum_{n=1}^N [\mu_n(A) + \mu_n(B)]$$

from which $\Phi(A+B) \le \Phi(A) + \Phi(B)$ by Theorem 1.16(b).

Corresponding to the spaces $\ell_p$ and $\ell_{p,w}$, we introduce the norm ideals $\mathcal{I}_p$ and $\mathcal{I}_{p,w}$. Before beginning their explicit study, we discuss two additional general properties of the $\mathcal{I}_\Phi$: Hölder's inequality and interpolation. In Chapter 3, we will discuss the duality theory for the $\mathcal{I}_\Phi$.

Hölder's inequality for the operator norms $\|\cdot\|_p$ has been around for many years (see, e.g., [**134**]) — its simplest proof is probably by complex interpolation (see [**251**, appendix to Section IX.4]). Using other interpolation methods, Simon [**296**] proved a Hölder inequality for the $\|\cdot\|_{p,w}$ operator norms. By mimicking the Goh'berg-Krein proof [**134**] of the $\|\cdot\|_p$ Hölder inequality, we are able to obtain a general result which includes the $\|\cdot\|_{p,w}$ result:

THEOREM 2.8 (Abstract Hölder Inequality). *Let $\Phi_1, \Phi_2, \Phi_3$ be three symmetric sequence norms with the property that if $a \in s_2$, $b \in s_3$, then $ab$, given by $(ab)_n = a_n b_n$, is in $s_1$ and*

$$\Phi_1(ab) \le \Phi_2(a)\Phi_3(b)$$

*Then, if $A \in \mathcal{I}_{\Phi_2}$, $B \in \mathcal{I}_{\Phi_3}$, it follows that $AB \in \mathcal{I}_{\Phi_1}$ and*

$$\Phi_1(AB) \le \Phi_2(A)\Phi_3(B)$$

*If either $A \in \mathcal{I}^{(0)}_{\Phi_2}$ or $B \in \mathcal{I}^{(0)}_{\Phi_3}$, then $AB \in \mathcal{I}^{(0)}_{\Phi_1}$. In particular, if*

$$p^{-1} = q^{-1} + r^{-1} \tag{2.5a}$$

*and $A \in \mathcal{I}_q$, $B \in \mathcal{I}_r$, then $AB \in \mathcal{I}_p$ and*

$$\|AB\|_p \le \|A\|_q \|B\|_r \tag{2.5b}$$

*and if $A \in \mathcal{I}_{q,w}$, $B \in \mathcal{I}_{r,w}$, $p > 1$, then $AB \in \mathcal{I}_{p,w}$ and*

$$\|AB\|_{p,w} \le \frac{p}{(p-1)} \|A\|_{q,w} \|B_{p,w} \tag{2.5c}$$

PROOF. By Horn's inequality, (1.12), and Corollary 1.10,

$$\sum_{n=1}^N \mu_n(AB) \le \sum_{n=1}^N \mu_n(A)\mu_n(B)$$

Then, using Theorem 1.9,

$$\begin{aligned}\Phi_1(AB) &= \Phi_1(\mu_n(AB)) \le \Phi_1(\mu_n(A)\mu_n(B)) \\ &\le \Phi_2(\mu_n(A))\Phi_3(\mu_n(B)) = \Phi_2(A)\Phi_3(B)\end{aligned}$$

(2.5b) then follows from the usual Hölder inequality and (2.5c) from the obvious inequalities

$$\left[\frac{(p-1)}{p}\right] \|ab\|_{p,w} \le \|a\|^*_{p,w} \|b\|^*_{r,w} \le \|a\|_{q,w} \|b\|_{r,w}$$ □

CONJECTURE. Let $\|A\|^*_{p,w} \equiv \|\mu_n(A)\|^*_{p,w} = \sup_n [n^{1/p}\mu_n(A)]$. Then

$$\|AB\|^*_{p,w} \le \|A\|^*_{q,w} \|B\|^*_{r,w}$$

The history of interpolation ideas in the context of operator ideals is somewhat murky. Kunze [**180**] introduced the complex method, including Riesz-Thorin and Stein theorems for $\mathcal{I}_p$ spaces, in the context of general non-commutative integration. The abstract complex interpolation theory of Calderón [**65**] and Lions [**205, 206**] is based on "wandering vector" theorems so that, given Kunze's work, such theorems were implicit in Calderón-Lions. Apparently unaware of this earlier work, Goh'berg and Krein [**134**] proved such theorems for $\mathcal{I}_p$ and a class of spaces including $\mathcal{I}_{p,w}$. Marcinkiewicz-type theorems were first discussed by Bennett [**29**]. The two abstract theorems below are essentially due to Simon [**296**].

Let $\Phi_0$ and $\Phi_1$ be two symmetric forms on finite sequences. Let $\Phi_t$ be the complex interpolated norms; we assume the reader is familiar with this notion (see, e.g., [**251**, appendix to Section IX.4]). It is not hard to see that each $\Phi_t$ is also a symmetric norm. We use $\mathcal{I}_{(t)}$ to stand for $\mathcal{I}_{\Phi_t}$.

THEOREM 2.9. *Let $\mathcal{I}_{(t)}$ be given as above. Let $z \mapsto A(z)$ be a map from the strip $0 \le \operatorname{Re} z \le 1$ to the bounded operators so that for any $\phi, \psi \in \mathcal{H}$, $(\phi, A(z)\psi)$ is continuous in the strip, analytic in its interior, and bounded. If $A(iy)$ (resp. $A(1+iy)$) lie in $\mathcal{I}_{(0)}$ (resp. $\mathcal{I}_{(1)}$) and*

$$M_t = \sup_y \Phi_t(A(t+iy)) < \infty$$

*for $t = 0, 1$, then $A(z) \in \mathcal{I}_{(\operatorname{Re} z)}$ for all $z$ in the strip and $M_t$ is convex.*

PROOF. As usual, by replacing $A(z)$ by $e^{a+bz}A(z)$, we can suppose that $M_0 = M_1 = 1$. Let $\phi, \psi \in \mathcal{B}$. Then $(\phi, A(z)\psi)_{n\le N}$ is an analytic $f$-valued function, so by definition of $\Phi_t$,

$$\begin{aligned} \Phi_t((\phi, A(t+iy)\psi)_{n\le N}) &\le \sup_{\substack{y \text{ real} \\ x=0 \text{ or } 1}} (\Phi_x((\phi, A(x+iy)\psi)_{n\le N}) \\ &\le \sup_{\substack{y \text{ real} \\ x=0 \text{ or } 1}} (\Phi_x(A(x+iy)) \le 1 \end{aligned}$$

by Proposition 2.6. Taking $N \to \infty$ and using Proposition 2.6 again, the theorem results if $\Phi_t$ is inequivalent to $\|\cdot\|_\infty$. The result for the case $\Phi_0 = \Phi_1 = \|\cdot\|_\infty$ requires an additional argument, which can be found in [**254**, p. 115]. □

THEOREM 2.10. *Let $\Phi_1, \Phi_2, \Phi_3$ be symmetric norms and $X_1, X_2, X_3$ Banach spaces with $X_1 \cap X_2$ dense in $X_3$.*

(a) *If every linear contraction from $X_1$ to $s_{\Phi_1}$ and $X_2$ to $s_{\Phi_2}$ is automatically a contraction from $X_3$ to $s_{\Phi_3}$, then the same is true if $s_\Phi$ is replaced by $\mathcal{I}_\Phi$.*
(b) *If every linear contraction from $s_{\Phi_1}$ to $X_1$ and $x_{\Phi_2}$ to $X_2$ is automatically a contraction from $s_{\Phi_3}$ to $X_3$, then the same is true if $s_\Phi$ is replaced by $\mathcal{I}_\Phi$.*

PROOF. (a) Let $T$ be a contraction from $X_i$ to $\mathcal{I}_{\Phi_i}$ $(i = 1, 2)$ and let $\phi, \psi \in \mathcal{B}$. Define $T_{\phi,\psi}(x) = \{\phi_n, T(x)\psi_n)\}$. Then, by Proposition 2.6, $T_{\phi,\psi}$ is a contraction from $X_i$ to $s_{\Phi_1}$ $(i = 1, 2)$ and so, by hypothesis, from $X_3$ to $s_{\Phi_3}$. Using Proposition 2.6 again, $T$ is a contraction from $X_3$ to $\mathcal{I}_{\Phi_3}$.

(b) Fix $A = \sum \mu_n(A)(\phi_n, \cdot)\psi_n$ in $\mathcal{I}_\infty$. Given a contraction from $\mathcal{I}_{\Phi_i}$ $(i = 1, 2)$ to $X_i$, let

$$S_{\phi,\psi}(\{a_n\}) = T\Big(\sum a_n(\phi_n, \cdot)\psi_n\Big)$$

Then $S_{\phi,\psi}$ is a contraction from $s_{\Phi_i}$ to $X_i$ $(i = 1, 2)$ and so from $s_{\Phi_3}$ to $X_3$. In particular,

$$\|T(A)\|_{X_3} = \|S_{\phi,\psi}(\mu_n(A))\|_{X_3} \le \Phi_3(\mu_n(A)) = \Phi_3(A)$$

□

REMARKS. 1. This theorem immediately implies Riesz-Thorin and Marcinkiewicz theorems for $\mathcal{I}_p$ and $\mathcal{I}_{p,w}$. While a Stein interpolation theorem is not explicitly included, it easily follows by the method of proof of Theorems 2.9 and 2.10.

2. Theorems 2.9, 2.10, and the proof of the triangle inequality in Theorem 2.7 show the power of Proposition 2.6. For $\{\mu_n(A)\}$ is a complicated non-linear function of $A$ while $(\phi_n, A\psi_n)$ is linear.

The remainder of these lectures concerns the theory and properties of $\mathcal{I}_p$ and $\mathcal{I}_{p,w}$. We next want to discuss some general criteria for an integral operator on $L^2(M, d\mu)$ to be in $\mathcal{I}_p$ (mainly for $p = 1, 2$). See also Addendum D. In Chapter 4, we return to this question and prove deep results for $(M, d\mu) = (\mathbb{R}^\nu, dx)$. (*Note:* $\mathcal{I}_1 \equiv$ "trace class"; $\mathcal{I}_2 \equiv$ "Hilbert-Schmidt operators.") The first result is classical:

THEOREM 2.11. *Let $\mathcal{H} = L^2(M, d\mu)$ for some separable measure space (i.e., one with $\mathcal{H}$ separable). If $A \in \mathcal{I}_2$, then there exists a unique function $K \in L^2(M \times M, d\mu \otimes d\mu)$ with*

$$(A\phi)(x) = \int K(x, y)\phi(y)\, d\mu(y) \tag{2.6}$$

*Conversely, any $K \in L^2(M \times M)$ defines an operator $A$ by* (2.6) *which is in $\mathcal{I}_2$ and*

$$\|A\|_2 = \|K\|_2 \tag{2.7}$$

REMARK. (2.6) is intended in the sense that the integral converges absolutely for a.e. $x$.

PROOF. Let $A \in \mathcal{I}_2$ and let $A = \sum \mu_n(A)(\phi_n, \cdot\,)\psi_n$ be its canonical expansion. The functions $\overline{\phi_n(y)}\,\psi_n(x)$ are orthonormal in $L^2(M \times M)$, so there is a function $K$ in $L^2$ obeying (2.7) so that

$$K = L^2 - \lim_{n\to\infty} \left( \sum_{n=1}^{N} \mu_n(A)\, \overline{\phi_n(y)}\, \psi_n(x) \right)$$

(2.6) is easily seen to hold.

Conversely, let $K \in L^2(M \times M)$. Then, by the Schwarz inequality,

$$\left| \int K(x, y)\phi(y)\, d\mu(y) \right| \le \left( \int |K(x, y)|^2\, d\mu(y) \right)^{1/2} \|\phi\|$$

so (2.6) defines a bounded operator. Moreover, given any $\phi, \psi \in \mathcal{B}$, $\overline{\phi_n}(x)\psi_n(y)$ are orthonormal in $L^2$, so

$$\sum |(\phi_n, A\psi_n)|^2 \le \|K\|_2^2$$

Thus, $A \in \mathcal{I}_2$ by Proposition 2.6 and $\|A\|_2 \le \|K\|_2$. This implies that the $K$ representing $A$ is unique and thus the equality in (2.7). □

Unfortunately, there is no simple necessary and sufficient condition for $A \in \mathcal{I}_1$ in terms of its integral kernel. Since we will not need the following, we quote it without proof (for $X = \mathbb{R}$, it is proven in Goh'berg-Krein [**134**, Section III.10] and for general $X$ in Reed-Simon [**253**, Section XI.4]):

THEOREM 2.12. *Let $\mu$ be a Baire measure on a locally compact space $X$. Let $K$ be a function on $X \times X$ which is continuous and Hermitian positive, that is, $\sum_{i=1}^{N} \bar{z}_i z_j K(x_i, x_j) \geq 0$ for any $x_1, \ldots, x_N \in X$, $z \in \mathbb{C}^N$, and any $N$. Then $K(x,x) \geq 0$ for all $x$. Suppose that, in addition,*

$$\int K(x,x)\, d\mu(x) < \infty$$

*Then there exists a unique operator $A$ in $\mathcal{J}_1$ obeying* (2.6) *and*

$$\|A\|_1 = \int K(x,x)\, d\mu(x)$$

REMARK. Continuity is clearly essential since (2.6) only determines $K$ a.e., and if $\mu$ has no pure points, $\{(x,x) \mid x \in X\}$ has $\mu \otimes \mu$ measure zero. Moreover, positivity is essential. For Carleman [**67**] has constructed a continuous real periodic function, $f$, on $[0, 2\pi]$ whose Fourier coefficients do not lie in any $\ell_p$ for $p < 2$. Thus $(Ag)(\theta) = \int f(\theta - \phi) g(\phi)\, d\phi$ defines an operator with $e^{in\theta}$ as eigenfunctions. Since $\hat{f} \notin \ell_p$, $p < 2$, $A \notin \mathcal{J}_p$ for $p < 2$. See Addendum D.

The following comparison theorem is sometimes useful:

DEFINITION. Let $A$ and $B$ be operators on $L^2(M, d\mu)$. We say that $B$ pointwise dominates $A$ if and only if

$$|(A\phi)(x)| \leq (B|\phi|)(x)$$

for all $\phi$.

THEOREM 2.13. *If $B$ pointwise dominates $A$ and $B \in \mathcal{J}_{2n}$ for some integer $n$, then $A \in \mathcal{J}_{2n}$ and $\|A\|_{2n} \leq \|B\|_{2n}$.*

PROOF. The intuition is based on the idea that if $A$ and $B$ have integral kernels $K$ and $L$, then domination says that $|K(x,y)| \leq L(x,y)$ and, by Theorem 2.11, $\|A\|_{2n}^{2n} = \int K(x_1, x_2) K^*(x_2, x_3) \ldots K^*(x_{2n}, x_1)$ where $K^*(x,y) = \overline{K(y,x)}$. Let $P$ be the projection onto a space spanned by the characteristic functions of a finite number of disjoint sets. Then $PAP$ is pointwise dominated by $PBP$ and each is given by an integral kernel so that $\|PAP\|_{2n} \leq \|PBP\|_{2n} \leq \|B\|_{2n}$. If one orders the $P$'s so that $P > Q$ if and only if $\operatorname{Ran} P \supset \operatorname{Ran} Q$, then $PAP \to A$ weakly as $P \to \infty$. Thus, by Fatou's lemma (Theorem 2.7(d)), $A \in \mathcal{J}_{2n}$ and $\|A\|_{2n} \leq \|B\|_{2n}$. □

CONJECTURE. Theorem 2.13 remains true if $\mathcal{J}_{2n}$ is replaced by $\mathcal{J}_p$ for any $p \geq 2$. This would imply the truth of a conjecture in [**296**]; namely, for any $f, g$ on $\mathbb{R}^\nu$, the operator with kernel $f(x-y)g(y)$ is in $\mathcal{J}_p$ $(p \geq 2)$ if $|f|(x-y)|g|(y)$ is in $\mathcal{J}_p$. We also note another conjecture in [**296**]: If $f^*$ is the symmetric decreasing rearrangement of $f$, then for $p \geq 2$, the $\mathcal{J}_p$ norm of $f^*(x-y)g^*(y)$ dominates the $\mathcal{J}_p$ norm of $f(x-y)g(y)$; for $p = 2n$, this is true by rearrangement inequalities [**56**]. The first conjecture here is now known to be false; see Addendum E.

REMARKS. 1. Theorem 2.13 is not true if $\mathcal{J}_{2n}$ is replaced by $\mathcal{J}_1$. For example, if $f$ is the function on $[0, 2\pi]$ which is 1 on $[0, \pi]$ and $-1$ on $[\pi, 2\pi]$, then $\phi \mapsto |f| * \phi$ is rank one and so in $\mathcal{J}_1$, while $\phi \mapsto f * \phi$ is *not* in $\mathcal{J}_1$ since the Fourier coefficients of $f$ go like $n^{-1}$. Presumably, Theorem 2.13 is not true if $\mathcal{J}_{2n}$ is replaced by any $\mathcal{J}_p$ with $p < 2$. Recently, Pitt [**239**] and Dodds-Fremlin [**98**] have proven the $p = \infty$ analog of Theorem 2.13.

2. The above examples show that proving an integral operator is in $\mathcal{J}_1$ can be quite difficult — this is unfortunate for $\mathcal{J}_1$ is often the natural space, for example, for defining $\det(1+A)$ and in the scattering theory of Chapter 6. However, the counterexamples which prevent nice theorems from holding are generally rather contrived, so that I have found the following to be true: If an integral operator with kernel $K$ occurs in some "natural" way and $\int |K(x,x)|\,dx < \infty$, then the operator can (almost always) be proven to be trace class (although sometimes only after some considerable effort).

Finally, we note the following criteria, the first of which is used as the basis of the definition of $\mathcal{J}_1$ in [**250**]:

THEOREM 2.14. *Let $A$ be a bounded, positive operator on $\mathcal{H}$ and let $\{\phi_n\}$ be an orthonormal basis. Then*

$$\sum_{n=1}^{\infty} (\phi_n, A\phi_n)$$

*is independent of the choice of basis. It is finite if and only if $A \in \mathcal{J}_1$ and, in that case, it equals $\|A\|_1$.*

THEOREM 2.15. *Let $B$ be a bounded operator on $\mathcal{H}$ and let $\{\phi_n\}$ and $\{\psi_n\}$ be orthonormal bases. Then*

$$\sum_{n=1}^{\infty} \|B\phi_n\|^2 \qquad \text{and} \qquad \sum_{n,m} |(\psi_m, B\phi_n)|^2$$

*are independent of the choice of bases and equal. They are finite if and only if $B \in \mathcal{J}_2$ and, in that case, they equal $\|B\|_2^2$.*

PROOFS. Choosing $B = \sqrt{A}$, Theorem 2.14 follows from Theorem 2.15. By Parseval's relation,

$$\sum_{n=1}^{\infty} \|B\phi_n\|^2 = \sum_{n,m} |(\psi_m, B\phi_n)|^2 = \sum_{n=1}^{\infty} \|B^*\psi_n\|^2$$

which proves the required equality and independence of basis. If $B \in \mathcal{J}_2$, then choosing the $\phi$'s to be a completion of the orthonormal set in $B = \sum_n \mu_n(B)(\phi_n, \cdot\,)\psi_n$, we see that the sum in question is finite and equal to $\sum \mu_n(B)^2 = \|B\|_2^2$. Conversely, suppose the sum is finite. Given any orthonormal sets $\phi, \psi$, one can complete them to a basis whence $\sum_n |(\phi_n, B\psi_n)|^2 \le \sum_{n,m} |\phi_n, B\psi_m)|^2$ so $B \in \mathcal{J}_2$ by Proposition 2.6. □

We close this chapter with a discussion of convergence theorems for $\mathcal{J}_p$ analogous to two celebrated theorems for $L^p$: the dominated convergence theorem and the theorem that weak (Banach space) convergence and convergence of the norms implies norm convergence. In some ways the proofs are simpler than for $L^p$; this is because $\mathcal{J}_p^*$ is closer to the discrete $\ell_p$ than the general $L^p$ space.

The non-commutative Fatou lemma (Theorem 2.7(d)) suggests that the analog of pointwise convergence should be weak operator convergence. The analog is seen to be good by considering $\ell_\infty$ embedded in $\mathcal{L}(\ell_2)$ as the family of diagonal matrices. Then $\mathcal{J}_p \cap \ell_\infty = \ell_p$ and a sequence $A^{(n)}$ with $\sup \|A^{(n)}\| < \infty$ converges to $A$ in the weak operator topology if and only if $a_m^{(n)} \to a_m$ for each $m$.

The following is clearly a direct analog of the dominated convergence theorem:

THEOREM 2.16. *Let $A_m, A, B$ be bounded operators on a Hilbert space, $\mathcal{H}$, with $B^* = B \geq 0$. Suppose that $|A_m| \leq B$ and $|A_m^*| \leq B$ for all $m$, $|A| \leq B$, $|A^*| \leq B$, and that $A_m \to A$ weakly. If $p < \infty$ and $B \in \mathcal{I}_p$, then $\|A - A_m\|_p \to 0$.*

PROOF. Fix $\varepsilon$. We can find a finite rank projection $P$ so that for $Q = 1 - P$, we have that

$$\|QBQ\|_p < \varepsilon$$

for example, by choosing $P$ to be the projection onto the span of the first few eigenvectors of $B$ in a canonical expansion. Thus, for $m$ large, $\|Q|A_m|Q\|_p \leq \varepsilon$ so $\||A_m|^{1/2}Q\|_{2p} \leq \varepsilon^{1/2}$ and similarly, $\||A_m^*|^{1/2}Q\|_{2p} \leq \varepsilon^{1/2}$. Using Hölder's inequality, $\|A_mQ\|_p \leq \||A_m|^{1/2}\|_{2p}\||A_m|^{1/2}Q\|_{2p} \leq \|B\|_p\varepsilon^{1/2}$ and similarly, $\|QA_mP\|_p \leq \|A_m^*Q\|_p \leq \|B\|_p\varepsilon^{1/2}$. Thus

$$\|A - A_m\|_p \leq 4\varepsilon^{1/2}\|B\|_p + \|P(A - A_m)P\|_p$$

Since $P$ is finite rank, $\|P(A - A_m)P\|_p \to 0$ as $m \to \infty$ by the weak convergence. Thus

$$\lim \|A - A_m\|_p \leq 4\varepsilon^{1/2}\|B\|_p$$

Since $\varepsilon$ is arbitrary, $\|A - A_m\|_p \to 0$. □

REMARK. If $\phi_n$ is an orthonormal set and $A_m = (\phi_1, \cdot)\phi_m$, then $A_m \to 0$ weakly and $|A_m| \leq (\phi_1, \cdot)\phi_1 \in \mathcal{I}_1$ but $\|A_m\| \nrightarrow 0$. This shows that the condition $|A_m^*| \leq B$ cannot be dropped.

In applications, this theorem is not as useful as its commutative analog for the following reason: Typically in the commutative case, one has $\|f_n - f_{n+1}\| \leq 2^{-n}$ and takes $g = |f_1| + \sum_{n=1}^{\infty}|f_{n+1} - f_n|$ whence $|f_m| \leq g$. Because the triangle inequality fails for $|\cdot|$ on operators, this trick will not work for $\mathcal{I}_p$ spaces. For this reason, the following device of Simon [**293**] is often useful.

DEFINITION. We write $A \underset{\mu}{<} B$ if $\mu_n(A) \leq \mu_n(B)$ for all $n$. We write $A \xrightarrow{\mu} B$ if $\mu_n(A) \to \mu_n(B)$ for all $n$.

Notice that $A \xrightarrow{\mu} 0$ if and only if $\|A\| \to 0$ since $\mu_n(A) \leq \mu_1(A) = \|A\|$. Moreover, by Theorem 1.20 (or a more elementary argument [**293**]), $|\mu_n(A) - \mu_n(B)| \leq \|A - B\|$ so that $A \xrightarrow{\mu} B$ if $\|A - B\| \to 0$.

THEOREM 2.17. *If $A_m \underset{\mu}{<} B$ with $B \in \mathcal{I}_p$ and if $\|A_m - A\| \to 0$, then $\|A_m - A\|_p \to 0$.*

PROOF. Since $\mu_n(A_m - A) \to 0$ by the above remark, it suffices to show that $\mu_n(A_m - A) \leq c_n$ for some $c \in \ell_p$, since we can then appeal to the dominated convergence theorem for $\ell_p$. As in the proof of Theorem 2.5, Fan's inequalities, (1.5) imply that

$$\mu_{2n+1}(A_m - A) \leq \mu_{n+1}(A) + \mu_{n+1}(A_m) \leq 2\mu_{n+1}(B)$$
$$\mu_{2n}(A_m - A) \leq 2\mu_n(B)$$

so we can take $c = (2\mu_1(B), 2\mu_1(B), 2\mu_2(B), \dots) \in \ell_p$. □

THEOREM 2.18 (Simon [**293**]). *Let $F$ be a continuous function on $[0,\infty)$ and let $\widetilde{F}(x) = \sup_{0<y<x}|F(y)|$. Suppose that $A_m \underset{\mu}{<} B$, $A_m \xrightarrow{\mu} A$, and that $\sum_n \widetilde{F}(\mu_n(B)) < \infty$. Then*

$$\sum_n F(\mu_n(A_m)) \to \sum_n F(\mu_n(A))$$

*as $m \to \infty$.*

PROOF. By continuity of $F$ and $A_m \xrightarrow{\mu} A$, we have that $\lim_{m\to\infty} F(\mu_n(A_m)) = F(\mu_n(A))$ for each $n$. But, by monotonicity of $\widetilde{F}$,

$$|F(\mu_n(A_m))| \leq \widetilde{F}(\mu_n(A_m)) \leq \widetilde{F}(\mu_n(B))$$

so the result follows from the dominated convergence theorem for $\ell_1$. □

The point of Theorem 2.18 is the following: Very often, one can prove inequalities for finite matrices; Theorem 2.18 provides the key for extensions of the theorem to infinite dimensions. Typically, $A_n = P_n A P_n$ with $P_n$ suitable finite rank projections.

The following result of Grümm [**145**] is often useful (it should be mentioned that, as we shall see, for $p \neq 1$, this result follows easily from the earlier theorems of McCarthy [**219**]).

THEOREM 2.19 (Grümm's Convergence Theorem). *Fix $p < \infty$. Suppose that $A_n \to A$ and $A_n^* \to A^*$ in the strong operator topology and that $\|A_n\|_p \to \|A\|_p$. Then $\|A_n - A\|_p \to 0$.*

PROOF. Since $\sup_n \|A_n\| < \infty$, we have that $A_n^* A_n \to A^* A$ strongly. Since $\sqrt{\cdot}$ is strongly continuous (we can uniformly approximate $f(x) = \sqrt{x}$ on $[0, \sup_n \|A_n\|]$ by polynomials), we see that $|A_n| \to |A|$ strongly. Similarly, $|A_n^*| \to |A^*|$ strongly. Thus, this theorem follows from the next. □

THEOREM 2.20. *Fix $p < \infty$. Suppose that $A_n \to A$, $|A_n| \to |A|$, and $|A_n^*| \to |A^*|$ all weakly, and that $\|A_n\|_p \to \|A\|_p$. Then $\|A_n - A\|_p \to 0$.*

PROOF. Without loss, we can suppose that $\|A\|_p = \|A_n\|_p = 1$. Fix $\varepsilon > 0$. Find a finite-dimensional projection $P$ so that $\|P|A|P\|_p \geq 1-\varepsilon$ and $\|P|A^*|P\|_p \geq 1-\varepsilon$; for example, $P$ can be the projection onto the span of the first few eigenvectors for $|A|$ and for $|A^*|$. Now, by the weak convergence, find $N$ so that for $n \geq N$, $\|P|A_n|P\| \geq 1 - 2\varepsilon$ and $\|P|A_n^*|P\| \geq 1 - 2\varepsilon$. By (1.20),

$$\|Q|A_n|Q\| \leq (1 - (1 - 2\varepsilon)^p)^{1/p}$$

and similarly for $\|Q|A_n^*|Q\|$. From this point, one need only follow the proof of Theorem 2.16. □

For $p > 1$, one can strengthen these results:

THEOREM 2.21. *Fix $1 < p < \infty$. Suppose that $A_n \to A$ weakly and that $\|A_n\|_p \to \|A\|_p$. Then $\|A - A_n\|_p \to 0$.*

FIRST PROOF. ($1 < p < \infty$) By the weak convergence, $\mathrm{Tr}(A_n B) \to \mathrm{Tr}(AB)$ for $B$ finite rank. Since $\|A_n\|_p$ is uniformly bounded, we can extend this to all $B \in \mathcal{I}_q$ with $q = p/(p-1)$. Thus $A_n \to A$ in the weak (Banach space) $\mathcal{I}_p$ topology. $\|A - A_p\| \to 0$ then follows from Theorems 1.23 and 1.24. □

SECOND PROOF. ($2 \le p < \infty$) Let $\eta_n$ be a basis of eigenvectors for $|A|$. Then for $m_0$ fixed,

$$\begin{aligned}
\|A\|_p^p &= \sum_{m=1}^{\infty} \|A\eta_m\|^p \\
&\le \sum_{\substack{m=1\\ m\neq m_0}}^{\infty} \liminf \|A_n\eta_n\|^p + \limsup \|A_n\eta_{m_0}\|^p \quad \text{(since } A_n\eta_m \to A\eta_m \text{ weakly)} \\
&\le \limsup \sum_{m=1}^{\infty} \|A_n\eta_m\|^p \quad \text{(by Fatou's lemma)} \\
&\le \limsup \|A_n\|_p^p \quad \text{(by (1.17) since } p \ge 2) \\
&= \|A\|_p^p \quad \text{(by hypothesis)}
\end{aligned}$$

Thus, $\|A\eta_{m_0}\| = \limsup \|A_n\eta_{m_0}\}$, so $A_n\eta_{m_0} \to A\eta_{m_0}$. Letting $m_0$ vary, we see that $A_n \to A$ strongly. Similarly, $A_n^* \to A^*$. Now use Grümm's convergence theorem. □

*Open question*: Does Theorem 2.21 hold for $p = 1$? This question has been settled; it does hold for $p = 1$. See Addendum H.

REMARKS. 1. All the above theorems on $\|A_n\|_p \to \|A\|_p$ fail for $p = \infty$. For example, let $\phi_n$ be an orthonormal set and let $A_n = (\phi_1, \cdot)\phi_1 + (\phi_n, \cdot)\phi_n$.

2. Clearly, by Grümm's theorem, $\|A_n - A\| \to 0$ and $\|A_n\|_p \to \|A\|_p$ imply $\|A_n - A\|_p \to 0$.

EXAMPLE 1. ($|\cdot|$ is continuous on each $\mathcal{J}_p$) Let $\|A_n - A\|_p \to 0$. Then $A_n^*A_n \to A^*A$ in operator norm so $|A_n| \to |A|$ in operator norm (by the argument in the proof of Theorem 2.19). Since $\||A_n|\|_p = \|A_n\|_p \to \|A\|_p = \||A_n|\|_p$, Grümm's theorem implies that $|A_n| \to |A|$ in $\|\cdot\|_p$.

EXAMPLE 2. (If $\alpha > 0$ and $p \ge \max(1, \alpha)$, then $A \to |A|^\alpha$ is continuous from $\mathcal{J}_p$ to $\mathcal{J}_{p/\alpha}$. This is a result of Brydges et al. [**64**].) As in Example 1, if $A_n \to A$ in $\mathcal{J}_p$, then $|A_n|^\alpha \to |A|^\alpha$ in operator norm. Since $\||A_n|^\alpha\|_{p/\alpha} = \|A_n\|_p^\alpha \to \|A\|_p^\alpha = \||A|^\alpha\|_{p/\alpha}$. Grümm's theorem completes the proof. Here is another proof, using instead Theorem 2.17: Without loss, we can suppose that $\|A_n - A_{n+1}\|_p \le 2^{-n}$. As noted just before Theorem 2.17, $\mu_m(A_n) \to \mu_m(A)$ so $|A_n|^\alpha \xrightarrow{\mu} |A|^\alpha$. It thus suffices to find $b_m$ in $\ell_{p/\alpha}$ with $\mu_m(|A_n|^\alpha) \le b_m$. Equivalently, we need only find $c_m$ in $\ell_p$ with $\mu_m(A_n) \le c_m$. By Theorem 1.20, $\|\mu.(A_n) - \mu.(A_{n+1})\|_{\ell_p} \le \|A_n - A_{n+1}\|_p \le 2^{-n}$, so we can take

$$c_m = \mu_m(A_1) + \sum_{n=1}^{\infty} |\mu_m(A_{n+1}) - \mu_m(A_n)|$$

EXAMPLE 3. (Grümm [**145**]) To illustrate how ad hoc methods can sometimes be useful, let us show that if $A_n \to A$ strongly, $\sup_n \|A_n\| < \infty$, and $\|B_n - B\|_p \to 0$, then $\|A_nB_n - AB\|_p \to 0$. For writing

$$\|A_nB_n - AB\|_p \le \|A_n\| \, \|B_n - B\|_p + \|(A_n - A)B\|_p$$

we see that it suffices to show that $\|(A_n - A)B\|_p \to 0$. Given $\varepsilon$, we can find $P$ finite rank so that $\|(1 - P)B\|_p < \varepsilon$. By the strong convergence, $\|(A_n - A)P\| \to 0$

so

$$\limsup \|(A_n - A)B\| \leq \varepsilon(\|A\| + \limsup \|A_n\|)$$

Since $\varepsilon$ is arbitrary, the result is proven. Note that the result is false if strong convergence is replaced by weak convergence (take $B_n = (\phi_1, \cdot\,)\phi_1$, $A_n = (\phi_1, \cdot\,)\phi_n$) or if $A_nB_n \to AB$ is replaced by $B_nA_n \to BA$ (take $B_n = (\phi_1, \cdot\,)\phi_1$ and $A_n = (\phi_n, \cdot\,)\phi_1$).

CHAPTER 3

# Trace, Determinant, and Lidskii's Theorem

Theorem 2.14 shows that any reasonable domain for the "trace" which is closed under $A \mapsto |A|$ must be contained in $\mathcal{J}_1$. On the other hand:

THEOREM 3.1. *Let $A = \sum \mu_n(A)(\phi_n, \cdot\,)\psi_n$ lie in $\mathcal{J}_1$. Then for any orthonormal basis $\{\eta_m\}$, $\sum_m |(\eta_m, A\eta_m)| < \infty$ and*

$$\sum_m (\eta_m, A\eta_m) = \sum_n \mu_n(A)(\phi_n, \psi_n) \equiv \mathrm{Tr}(A)$$

*is independent of basis. Moreover,*

$$|\mathrm{Tr}(A)| \leq \|A\|_1 \tag{3.1}$$

*$A \mapsto \mathrm{Tr}(A)$ is a bounded linear functional on $\mathcal{J}_1$, and for $A \in \mathcal{J}_1$ and $B \in \mathcal{L}(\mathcal{H})$, $\mathrm{Tr}(AB) = \mathrm{Tr}(BA)$.*

PROOF. Let $a_{nm} = (\phi_n, \eta_m)(\eta_m, \psi_n)$. As usual, $\sum_m |a_{nm}| \leq 1$ for each $n$ so that $\sum_m |(\eta_m, A\eta_m)| \leq \sum_{n,m} |a_{nm}| \mu_n(A) \leq \|A\|_1$, incidentally proving (3.1). The absolute convergence of the last double sum justifies an interchange in

$$\sum_m (\eta_m, A\eta_m) = \sum_{n,m} a_{nm}\mu_n(A) = \sum_n \mu_n(A) \sum_m a_{nm} = \sum_n \mu_n(A)(\phi_n, \psi_n)$$

by the Parseval relation. Linearity of $A \mapsto \mathrm{Tr}(A)$ follows from the absolute convergence of the sums. If $A = \sum \mu_n(A)(\phi_n, \cdot\,)\psi_n \in \mathcal{J}_1$ and $B$ is bounded, then

$$\mathrm{Tr}(AB) = \sum_n (\psi_n, AB\psi_n) = \sum_n \mu_n(A)(\phi_n, B\psi_n) = \sum_n (\phi_n, BA\phi_n) = \mathrm{Tr}(BA) \quad \square$$

One point of the trace is the duality theory for the $\mathcal{J}_\Phi$ which results.

THEOREM 3.2 (Schatten [**278**]). *Let $\Phi$ and $\Phi'$ be conjugate symmetric norms. Then $A \in \mathcal{J}_\Phi$ and $B \in \mathcal{J}_{\Phi'}$ imply that $AB \in \mathcal{J}_1$. Moreover, for each fixed $B \in \mathcal{J}_{\Phi'}$, the map $A \mapsto \mathrm{Tr}(AB)$ is a bounded linear functional in $\mathcal{J}_\Phi$ with norm $\Phi'(B)$. If $\Phi'$ is inequivalent to $\ell_\infty$, every such functional on $\mathcal{J}_\Phi^{(0)}$ is of this form, that is, $(\mathcal{J}_\Phi^{(0)})^* = \mathcal{J}_{\Phi'}$ (isometrically). In particular, $\mathcal{J}_\Phi$ is reflexive if and only if both $\Phi$ and $\Phi'$ are regular. If $\Phi = \ell_1$, then $\mathcal{J}_\Phi^* = \mathcal{L}(\mathcal{H})$ with its norm.*

PROOF. By definition of conjugate, $\sum |a_n b_n| \leq \Phi(a)\Phi'(b)$, so $AB \in \mathcal{J}_1$ if $A \in \mathcal{J}_\Phi$ and $B \in \mathcal{J}_{\Phi'}$ on account of Hölder's inequality (Theorem 2.8) and $|\mathrm{Tr}(AB)| \leq \|AB\|_1 \leq \Phi(A)\Phi'(B)$ so that $A \mapsto \mathrm{Tr}(AB)$ defines a linear functional $\ell_B$ of norm bounded by $\Phi'(B)$. On the other hand, if $B = \sum \mu_n(B)(\phi_n, \cdot\,)\psi_n$, and we consider $A$'s of the form $\sum_{n=1}^N a_n(\psi_n, \cdot\,)\phi_n$, we see that $\|\ell_B\| \geq \Phi'(\mu_1(B), \dots, \mu_N(B), 0, \dots)$ so that $\|\ell_B\| = \Phi'(B)$.

Now let $\ell$ be a bounded linear function on $\mathcal{I}_\Phi^{(0)}$. Then $\phi, \psi \mapsto \ell((\phi, \cdot)\psi)$ defines a bounded Hermitian bilinear map on $\mathcal{H}$, so we can find $B$ with

$$\operatorname{Tr}(B(\phi, \cdot)\psi) \equiv (\phi, B\psi) = \ell((\phi, \cdot)\psi)$$

If we show that $B \in \mathcal{I}_{\Phi'}$ (resp. $\mathcal{L}(\mathcal{H})$ for $\Phi' = \ell_\infty$), then, by continuity, $\ell(A) = \operatorname{Tr}(AB)$ for any $A$ in the closure of the finite rank operators, that is, $\mathcal{I}_\Phi^{(0)}$.

Fix $\{\phi_n\}_{n=1}^N$, $\{\psi_n\}_{n=1}^N$ finite orthonormal sets. Then $(a_1, \dots, a_N) \mapsto \ell(\sum a_n(\phi_n, \cdot)\psi_n) = \sum a_n(\phi_n, B\psi_n)$ is a bounded map in $\Phi$ norm so $\Phi'((\phi_n, B\psi_n)_{n\leq N}) \leq \|\ell\|$. Taking $N$ to infinity and using Proposition 2.6, we see that $B \in \mathcal{I}_{\Phi'}$ in case $\Phi'$ is inequivalent to $\ell_\infty$. $\square$

Once the trace is defined, a natural question arises, which we will call the Lidskii problem in honor of its solver. Let $A \in \mathcal{I}_1$ and let $\{\lambda_n(A)\}$ be its eigenvalues counted up to algebraic multiplicity. By Weyl's inequality, Theorem 1.15, $\sum_n \lambda_n(A)$ converges absolutely. Is it true that

$$\sum_n \lambda_n(A) = \operatorname{Tr}(A) \tag{3.2}$$

The answer is yes, as we will see. The proof is surprisingly difficult, only being accomplished in 1959. To understand why it is difficult, consider an operator $A$ which is quasi-nilpotent, that is, $\|A^n\|^{1/n} \to 0$ as $n \to \infty$. It follows that for such an operator, each $\lambda_n(A)$ is zero. Why should the sum of diagonal matrix elements be zero? Moreover, the equality (3.2) is false in a general Banach space although there is a natural trace class and a natural trace; see Chapter 10.

How does one prove (3.2) in the finite-dimensional case? There are two natural proofs: One goes through the construction of a Jordan normal form, after which it is obvious; the other notes that $\operatorname{Tr}(A)$ is one of the coefficients in the characteristic polynomial $\det(A - \lambda)$, precisely the coefficient which is the sum of the roots. Lidskii's original 1959 proof [**194**] essentially follows the first strategy — using an approximation argument for the quasi-nilpotent case and a Jordan normal form on the non-zero eigenvalues. The proof which we prefer follows Simon [**300**] and develops characteristic polynomials. In their second volume, Dunford and Schwartz [**105**] give an intermediate proof: The non-zero eigenvalues are treated by a Jordan normal form; the quasi-nilpotent part by a characteristic polynomial — the rather subtle Hadamard factorization we need is then replaced by Liouville's theorem! Interestingly enough, Carleman, in a famous 1921 paper [**68**], proved that

$$\operatorname{Tr}(A^2) = \sum \lambda_n(A)^2 \tag{3.3}$$

for $A$ in the Hilbert-Schmidt class, $\mathcal{I}_2$. (3.3) is implied by (3.2) but not conversely, even in the finite-dimensional case. Actually, Carleman just dealt with integral operators on $L^2(a, b)$ with square integrable kernel — I don't believe it was known at the time that this was somehow independent of the realization as $L^2(a, b)$. Moreover, he never explicitly proved (3.3). However, he considers a certain function $\det_2(1 + \mu A)$ (the same $\det_2$ we will discuss in Chapters 4 and 9) given by an explicit power series expansion

$$\det_2(1 + \mu A) = 1 - \tfrac{1}{2}\mu^2 \operatorname{Tr}(A^2) + \cdots$$

The last result of his paper is an absolutely convergent product expansion

$$\det_2(1+\mu A) = \prod_n (1+\mu\lambda_n(A))e^{-\mu\lambda_n(A)}$$

from which (3.3) follows by equating the coefficients of the $\mu^2$ terms.

A solution of the Lidskii problem via a characteristic polynomial will require us to develop the determinant, an extremely useful function in any event. There are a variety of definitions of $\det(1+A)$ for $A \in \mathcal{J}_1$. Besides the one we use, one going back to Fredholm [**115**] (in a modern guise advocated especially by Grothendieck [**144**]), Goh'berg-Krein [**134**] use $\det(1+A) = \prod_n (1+\lambda_n(A))$, and Dunford-Schwartz [**105**] $\det(1+\mu A) = e^{\mathrm{Tr}(\log(1+\mu(A))}$ for $\mu$ small followed by analytic continuation. If you return to these definitions after the presentation below, you will see that their equivalence is precisely Lidskii's theorem, that is, the equality (3.2).

To motivate our definition, let $A$ be an operator on $\mathbb{C}^N$. Let $\lambda_1(A),\dots,\lambda_N(A)$ be the eigenvalues of $A$. Then $\lambda_{i_1}(A)\dots\lambda_{i_k}(A)$ $(i_1<\dots<i_k)$ are the eigenvalues of $\Lambda^k(A)$, so

$$\mathrm{Tr}(\Lambda^k(A)) = \sum_{i_1<\dots<i_k} \lambda_{i_1}(A)\dots\lambda_{i_k}(A)$$

and so

$$\sum_{k=0}^N \mathrm{Tr}(\Lambda^k(A)) = \prod_{n=1}^N (1+\lambda_n(A))$$

LEMMA 3.3. *If $A$ is a trace class operator on a Hilbert space $\mathcal{H}$, then $\Lambda^k(A)$ is a trace class operator on $\Lambda^k\mathcal{H}$ with*

$$\|\Lambda^k(A))\|_1 \le (k!)^{-1}\|A\|_1^k \tag{3.4}$$

*In particular, the series*

$$\det(1+zA) \equiv \sum_{k=0}^\infty z^k \mathrm{Tr}(\Lambda^k(A)) \tag{3.5}$$

*defines an entire function obeying*

$$|\det(1+zA)| \le \exp(|z|\,\|A\|_1) \tag{3.6}$$

*Moreover, for any fixed $\varepsilon>0$ and $A$ in $\mathcal{J}_1$, we have that*

$$|\det(1+zA)| \le C_\varepsilon \exp(\varepsilon|z|) \tag{3.7}$$

PROOF. $|\Lambda^k(A)| = \Lambda^k(|A|)$ so $\Lambda^k(A)$ has singular values $\mu_{i_1}(A)\dots\mu_{i_k}(A)$ with $i_1<\dots<i_k$, so

$$\|\Lambda^k(A)\|_1 = \sum_{i_1<\dots<i_k} \mu_{i_1}(A)\dots\mu_{i_k}(A) \tag{3.8}$$

is finite and (3.4) holds. From (3.4), we obtain (3.5) and (3.6) immediately. Moreover, (3.8) implies that

$$|\det(1+zA)| \le \prod_{n=1}^\infty (1+|z|\mu_n(A))$$

Choosing $N$ so $\sum_{n\ge N+1}\mu_n(A) \le \varepsilon/2$ and using $(1+x)\le e^x$, we see that

$$|\det(1+zA)| \le \prod_{n=1}^N (1+|z|\mu_n(A))e^{|z|\varepsilon/2} \le C_\varepsilon e^{\varepsilon|z|}$$

since exponentials bound polynomials. □

THEOREM 3.4. *$A \mapsto \det(1+A)$ is a continuous function on $\mathcal{I}_1$. Explicitly,*

$$|\det(1+A) - \det(1+B)| \le \|A-B\|_1 \exp(\|A\|_1 + \|B\|_1 + 1) \tag{3.9}$$

PROOF. Let $F(A) = \det(1+A)$. We first note that $g(z) = F(C+zD)$ is analytic in $z$ since $\operatorname{Tr}(\Lambda^k(C+zD))$ is a polynomial of degree $k$ in $z$. We now use an argument of Seiler-Simon [**288**]: Let $g(z) = F(\frac{1}{2}(A+B) + z(A-B))$. Then

$$\begin{aligned}\|F(A)-F(B)\|_1 & \\ &= \|g(\tfrac{1}{2}) - g(-\tfrac{1}{2})\|_1 \\ &\le \sup_{-1/2\le t\le 1/2} \|g'(t)\|_1 \\ &\le R^{-1} \sup_{|z|\le R+1/2} \|g(z)\|_1 \\ &\le \|A-B\|_1 \exp(\|\tfrac{1}{2}(A+B)\|_1 + \tfrac{1}{2}\|A-B\|_1 + \|A-B\|_1^{-1}\|A-B\|_1)\end{aligned}$$

where in the last step, we pick $R = (\|A-B\|_1)^{-1}$ and use (3.6). In the next to last step, we use a Cauchy estimate. □

THEOREM 3.5. (a) *For any $A$ and $B$ in $\mathcal{I}_1$,*

$$\det(1+A+B+AB) = \det(1+A)\det(1+B) \tag{3.10}$$

(b) *For $A \in \mathcal{I}_1$, $\det(1+A) \neq 0$ if and only if $1+A$ is invertible.*
(c) *For any $A \in \mathcal{I}_1$ and $z_0 = -\lambda^{-1}$ with $\lambda$ an eigenvalue of algebraic multiplicity $n$, $\det(1+zA)$ has a zero of order $n$ at $z_0$.*

PROOF. (a) By the continuity (3.9), we need only prove (3.10) for finite rank operators where it is essentially just the fact that, for finite-dimensional operators, $\det(CD) = \det(C)\det(D)$.

(b) If $1+A$ is not invertible, $\det(1+A) = 0$ by part (c). If $1+A$ is invertible, let $B = -A(1+A)^{-1}$. Then $(1+A+B+AB) = 1$, so $\det(1+A) \neq 0$ by (3.10).

(c) Let $P$ be the spectral projection for $\lambda^{-1}$ as discussed in Section 1.2. Then, by $((PA))(1-P)A = 0$ and (3.10), we have

$$\det(1+zA) = \det(1+zAP)\det(1+zA(1-P))$$

$1 + z_0A(1-P)$ is invertible since $\lambda \notin \sigma(A(1-P))$, so it suffices to show that $\det(1+zAP)$ has an $n$-th order zero at $z_0$. But $AP$ is a finite rank operator with eigenvalue $\lambda$ of multiplicity $n$, so $\det(1+zAP) = (1-zz_0^{-1})^n$ by the evaluation of a finite-dimensional determinant. □

The final element of our solution of the Lidskii problem is the following subtle Hadamard product formula (see Boas [**48**], Titchmarsh [**329**], or Reed-Simon [**254**] for the proof):

LEMMA 3.6. *Let $f(z)$ be an entire function with zeros at $z_1, z_2, \dots$ (counting multiplicity). Suppose that $f(0) = 1$, $\sum_{n=0}^\infty |z_n|^{-1} < \infty$, and that for any $\varepsilon$, $|f(z)| \le C_\varepsilon \exp(\varepsilon|z|)$. Then*

$$f(z) = \prod_{n=1}^\infty (1 - z_n^{-1}z)$$

THEOREM 3.7. *For any* $A \in \mathcal{J}_1$,

$$\det(1+zA) = \prod_n (1+z\lambda_n(A)) \tag{3.11}$$

*In particular, Lidskii's equality,* (3.2), *holds.*

PROOF. Let $f(z) = \det(1+zA)$. By Theorem 3.5, it has zeros at points $z_n = -\lambda_n^{-1}(A)$ and by the Lalesco-Schur-Weyl inequality (1.15) for $p=1$, $\sum |z_n|^{-1} < \infty$. Clearly, $f(0) = 1$ and, by Lemma 3.3, $f(z) \leq C_\varepsilon \exp(\varepsilon|z|)$. Thus, by Lemma 3.6, the expansion (3.11) is valid. The right side is clearly $1 + z(\sum \lambda_n) + 0(z^2)$, proving (3.2). □

REMARK. One might hope that the higher identities on Taylor coefficients in (3.11) would yield more information, but the $m$-th one is just Lidskii's equality, (3.2), for $\Lambda^m(A)$!

COROLLARY 3.8. *If* $A, B \in \mathcal{L}(\mathcal{H})$ *have the property that both* $AB$ *and* $BA$ *lie in* $\mathcal{J}_1$ *(in particular, if* $A \in \mathcal{J}_\Phi$, $B \in \mathcal{J}_{\Phi'}$*), then* $\operatorname{Tr}(AB) = \operatorname{Tr}(BA)$.

PROOF. By a well-known result (see, e.g., [**87**]), $AB$ and $BA$ have the same non-zero eigenvalues, including identical algebraic multiplicities, that is, $\lambda_n(AB) = \lambda_n(BA)$. Lidskii's equality, (3.2), completes the proof. □

Finally, we want to give explicit formulae for $\operatorname{Tr}(A)$ and $\det(A)$ for some simple integral operators. For simplicity, we suppose that

$$(Af)(x) = \int_a^b K(x,y) f(y)\, dy \tag{3.12}$$

on $L^2(a,b)$, with $-\infty < a < b < \infty$ and with $K$ continuous on $[a,b] \times [a,b]$.

THEOREM 3.9. *Let* $A$ *in* $\mathcal{J}_1$ *have the form* (3.12) *with* $K$ *continuous. Then*

$$\operatorname{Tr}(A) = \int_a^b K(x,x)\, dx \tag{3.13}$$

PROOF. For each $n$, let $\{\phi_{n;m}\}_{m=1}^{2^n}$ be the functions

$$\phi_{n;m}(x) = \begin{cases} 2^{n/2} & \frac{m-1}{2^n} \leq x < \frac{m}{2^n} \\ 0 & \text{otherwise} \end{cases}$$

Let $P_n$ be the projection onto the span of the $\phi_{n;m}$. Then $\operatorname{Tr}(A) = \lim_{n\to\infty} \operatorname{Tr}(P_n A P_n)$ since one can easily construct a basis $\psi_1, \dots$ for $L^2(a,b)$ with $\psi_1, \dots, \psi_{2^n} \in \operatorname{Ran} P_n$. Now

$$\begin{aligned} \operatorname{Tr}(P_n A P_n) &= \sum_m (\phi_{n;m}, A\phi_{n;m}) \\ &= 2^n \sum_{m=1}^n \iint_{(m-1)/2^n \leq x,\, y < m/2^n} [K(x,y)]\, dx\, dy \end{aligned}$$

Using the uniform continuity of $K$, one easily proves that this integral converges to $\int_a^b K(x,x)\, dx$ as $n \to \infty$. □

THEOREM 3.10. *Let $A$ in $\mathcal{J}_1$ have the form* (3.12) *with $K$ continuous. Then*

$$\det(1+A) = \sum_{n=0}^{\infty} \frac{\alpha_n}{n!} \tag{3.14a}$$

*where*

$$\alpha_n = \int_a^b \cdots \int_a^b dx_1 \dots dx_n \, K\begin{pmatrix} x_1 & \dots & x_n \\ x_1 & \dots & x_n \end{pmatrix} \tag{3.14b}$$

*and*

$$K\begin{pmatrix} x_1 & \dots & x_n \\ y_1 & \dots & y_n \end{pmatrix} = \det[(K(x_i, y_j))_{1 \le i, j \le n}] \tag{3.14c}$$

REMARK. (3.14) is the original expansion of Fredholm [**115**]; we call it the Fredholm formula.

PROOF. Let $Q_n$ be the projection from $\otimes^n L^2(a,b)$ onto $\Lambda^n(L^2(a,b))$ and let $C_n = Q_n \Lambda^n(A) Q_n$. We claim that $C_n$ is an integral operator with kernel

$$(n!)^{-1} K\begin{pmatrix} x_1 & \dots & x_n \\ y_1 & \dots & y_n \end{pmatrix}$$

whence its trace is given by $(n!)^{-1}\alpha_n$ (as in the proof of Theorem 3.10), so (3.14) holds. To check the kernel, we note that $C_n$ is determined by

$$\begin{aligned} &C_n(f_1 \otimes \cdots \otimes f_n) \\ &\quad = (n!)^{-1} \sum_{\pi \in \sigma_n} (-1)^\pi A f_{\pi(1)} \otimes \cdots \otimes A f_{\pi(n)} \\ &\quad = (n!)^{-1} \sum_{\pi} \int \cdots \int (-1)^\pi K(x_1, y_{\pi(1)}) \dots K(x_n, y_{\pi(n)}) f_1(y_1) \dots f_n(y_n) \\ &\quad = (n!)^{-1} \int \cdots \int K\begin{pmatrix} x_1 & \dots & x_n \\ y_1 & \dots & y_n \end{pmatrix} f_1(y_1) \dots f_n(y_n) \end{aligned}$$

□

CHAPTER 4

# $f(x)g(-i\nabla)$

In this chapter, we want to consider operators on $L^2(\mathbb{R}^\nu)$ formally given as $f(x)g(-i\nabla)$. (For operators on $L^2(X,d\rho)$, $X\subset\mathbb{R}^\nu$, see [**37, 38, 40, 44**]). If $f$ and $g$ are a.e. finite measurable function on $\mathbb{R}^\nu$, the sets $\mathcal{D}_f$ and $\mathcal{D}_g$ in $L^2$ given by $\mathcal{D}_f=\{h\in L^2\mid fh\in L^2\}$, $\mathcal{D}_g=\{h\in L^2\mid gh\in L^2\}$ are dense in $L^2$. For $\phi\in\mathcal{D}_f$ and $\widehat{\psi}\in\mathcal{D}_g$, ($\widehat{\ }\equiv$ Fourier transform), we can unambiguously define $(\phi,A\psi)$ with $A$ given formally by $f(x)g(-i\nabla)$ as the inner product of $\bar f\phi$ and $(g\widehat{\psi})^\vee$ ($^\vee\equiv$ inverse Fourier transform). Some of the most celebrated estimates in analysis assert that $A$ is a bounded operator (to be pedantic, there is a bounded operator $\tilde A$ with $(\phi,A\psi)=(\phi,\tilde A\psi)$ for $\phi\in\mathcal{D}_f$, $\tilde\psi\in\mathcal{D}_g$; henceforth, we will be sloppy about this point). For example,

(a) *Young's inequality.* If $g\in\mathcal{S}'$, then $(g\widehat{\psi})^\vee=h*\psi$ where $h=(2\pi)^{-\nu/2}\widehat{g}$. Suppose that $h\in L^{p'}$, $f\in L^p$ with $2\le p\le\infty$ (and $p'=p/(p-1)$). Then Hölder's and Young's inequalities assert that

$$\|f(h*\phi)\|_2\le\|f\|_p\|h\|_{p'}\|\phi\|_2 \tag{4.1}$$

Moreover, (4.1) implies that $h*$ is bounded from $L^2$ to $L^r$ where $r^{-1}=\frac12-p^{-1}$.

(b) *Sobolev inequalities.* Recall [**319**] that $L^p_w(\mathbb{R}^\nu)$ is defined to be the set of $f$'s obeying $|m_f(t)|\le(\|f\|^*_{p,w}t^{-1})^p$ where $m_f(t)\equiv|\{x\mid|f(x)|\ge t\}|$ (with $|\cdot|=$ Lebesgue measure). Sobolev's inequalities imply that $h*$ is bounded from $L^2$ to $L^r$ if $r^{-1}=\frac12-p^{-1}$ with $h\in L^{p'}_w$ and $2<p<\infty$. (This is proven, e.g., in [**251**, Section IX.4]; the relation to the more usual Sobolev inequalities involving gradients comes from the fact that inverses of differential operators tend to be convolutions by functions in weak spaces, e.g., for $\nu\ge3$, $(-\Delta)^{-1}$ is convolution by $c_\nu|x|^{-\nu+2}$.) Thus, using Hölder's inequality ($2<p<\infty$),

$$\|f(h*\phi)\|_2\le C_p\|f\|_p\|h\|^*_{p',w}\|\phi\|_2 \tag{4.2}$$

and (4.2) implies Sobolev's inequalities from $L^2$ to $L^r$.

(c) *Strichartz inequalities* [**323**]. Strichartz has extended (4.2) to

$$\|f(h*\phi)\|_1\le\tilde C_p\|f\|^*_{p,w}\|h\|^*_{p',w}\|\phi\|_2 \tag{4.3}$$

It is natural to ask if $\phi\mapsto f(h*\phi)$ has nice trace ideal properties. The Hausdorff-Young inequalities assert that for $2\le p\le\infty$, $\check g\in L^{p'}$ implies that $g\in L^p$, so the results below answer this question in cases (a) and (b).

THEOREM 4.1. *If $f\in L^p$ and $g\in L^p$ with $2\le p<\infty$, then $f(x)g(-i\nabla)$ is in $\mathcal{J}_p$ and*

$$\|f(x)g(-i\nabla)\|_p\le(2\pi)^{-\nu/p}\|f\|_p\|g\|_p \tag{4.4}$$

THEOREM 4.2 (Cwikel [**80**]; conjectured in Simon [**297**]). *If $f \in L^p$ and $g \in L^p_w$ with $2 < p < \infty$, then $f(x)g(-i\nabla)$ is in $\mathcal{I}_{p,w}$ and*

$$\|f(x)g(-i\nabla)\|_{p,w} \le C_p \|f\|_p \|g\|^*_{p,w} \tag{4.5}$$

We defer the proof of these results. As for (4.3), we note

PROPOSITION 4.3. *Given $p \in (2,\infty)$, there exist $f, g \in L^p_w$ so that $f(x)g(-i\nabla)$ is not compact.*

PROOF. If $f(x) = |x|^{-\nu/p}$, $g(y) = |y|^{-\nu/p}$, then $A = f(x)g(-i\nabla)$ commutes with the unitary group of dilations,

$$(U_\lambda \phi)(x) = e^{\lambda/2}\phi(e^\lambda x)$$

Since $U$ has purely continuous spectrum, $A$ cannot be compact. □

We will also prove later that (4.4) is best possible at the endpoint $p = 2$:

PROPOSITION 4.4. *If $f, g$ are both non-zero and $f(x)g(-i\nabla) \in \mathcal{I}_2$, then $f$ and $g$ both lie in $L^2$.*

The above leaves open the important question of when such operators lie in $\mathcal{I}_1$. To answer this, it is useful to introduce a class of spaces following Birman-Solomjak [**40**]:

DEFINITION. For $\alpha \in \mathbb{Z}^\nu$, let $\Delta_\alpha$ be the unit cube in $\mathbb{R}^\nu$ with center at $\alpha$ and let $\chi_\alpha$ be its characteristic function. $\ell^q(L^p)$ is the set of functions for which

$$\|f\|_{p;q} = \left(\sum_\alpha \|f\chi_\alpha\|_p^q\right)^{1/q}$$

is finite where $\|g\|_p$ is the usual $L^p(\mathbb{R}^\nu)$ norm. For example,

$$\|f\|_{2;1} = \sum_\alpha \left(\int_{x\in\Delta_\alpha} |f(x)|^2\, dx\right)^{1/2}$$

If $a(\alpha) \equiv \|f\chi_\alpha\|_p$ is in weak $\ell^q$, we say $f \in \ell^q_w(L^p)$ and write $\|f\|^*_{p;q,w} = \|a\|^*_{q,w}$.

THEOREM 4.5 (Birman-Solomjak [**40, 45**]). *If $f$ and $g$ both lie in $\ell^p(L^2)$ for $1 \le p \le 2$, then $f(x)g(-i\nabla)$ is in $\mathcal{I}_p$ and*

$$\|f(x)g(-i\nabla)\|_p \le C_p \|f\|_{2;p} \|g\|_{2;p} \tag{4.6}$$

The following result seems to be new, although it is a simple combination of ideas of Birman-Solomjak [**40**] and Cwikel [**80**]:

THEOREM 4.6. *Let $1 < p < 2$. If $f \in \ell^p(L^2)$ and $g \in \ell^p_w(L^2)$, then $f(x)g(-i\nabla) \in \mathcal{I}_{p,w}$ and*

$$\|f(x)g(-i\nabla)\|_{p,w} \le C_p \|f\|_{2;p} \|g\|^*_{2;p,w}$$

As in the $p \ge 2$ case, the $\mathcal{I}_p$ theorem is sharp at the endpoint, showing the naturalness of $\ell^1(L^2)$ conditions.

PROPOSITION 4.7. *If $f, g$ are both non-zero and $f(x)g(-i\nabla) \in \mathcal{I}_1$, then $f$ and $g$ both lie in $\ell^1(L^2)$.*

Before turning to the proof of these results, we make some remarks about $\ell^1(L^2)$:

REMARKS. (a) Since $(\int_{\Delta_\alpha} |f(x)|^2\,dx)^{1/2} \geq \int_{\Delta_\alpha} |f(x)|\,dx$, the $\ell^1(L^2)$ norm dominates the $L^1$ norm. Since $(\sum a_n)^2 \geq \sum a_n^2$ for $a_n \geq 0$, the $\ell^1(L^2)$ norm dominates the $L^2$ norm. Thus $\ell^1(L^2) \subset L^1 \cap L^2$.

(b) Let $\Delta_n$ be some sequence of disjoint cubes and let $f_n$ be the characteristic function of a set of measure $n^{-2}$ in $\Delta_n$. Then $f = \sum f_n$ is in $L^1 \cap L^\infty$ but not in $\ell^1(L^2)$.

(c) Let $L_\delta^2 = \{f \mid (1+x^2)^{\delta/2} f \in L\}$ with the obvious norm. Then $\ell^1(L^2) \supset L_\delta^2$ for $\delta > \nu/2$, for

$$\|f\chi_\alpha\|_2 \leq \left[\left(1+\left(\frac{\alpha}{2}\right)^2\right)^{-\delta/2}\right] \|\chi_\alpha(1+x^2)^{\delta/2} f\|_2 \equiv a(\alpha)b(\alpha)$$

Since $\delta > \nu/2$, $\sum|a(\alpha)|^2 < \infty$ and since $\sum|b(\alpha)|^2 = \|f\|_{L_\delta^2}^2$, we have $\|f\|_{2;1} \leq C\|f\|_{L_\delta^2}$ by the Schwarz inequality.

In particular, (c) and Theorem 4.5 imply:

COROLLARY 4.8. *If $f, g \in L_\delta^2$, then $f(x)g(-i\nabla) \in \mathcal{I}_1$.*

(d) As we shall see, Theorem 4.5 follows from Corollary 4.8, indeed, from the special case where $f$ and $g$ lie in $L^2(\Delta_{\alpha=0})$! The earliest trace class result of this genre was actually very close to being sufficient for this fact: namely, Stinespring [**320**] showed that $f(x)g(-i\nabla) \in \mathcal{I}_1$ if $f \in L_\delta^2$ and $g \in L^\infty(\Delta_{\alpha=0})$. Birman-Solomjak [**40**], after introducing their space, proved the result for $f \in \ell^1(L^2)$, but required $g \in L_\delta^2$. The full result is in [**45**]. Unaware of this work, Corollary 4.8 was rediscovered by Kato [**166**] and Reed-Simon [**252**]. Proposition 4.7 appears to be new.

(e) At the conclusion of this chapter, we give a result for $f(x)g(-i\nabla)h(z)$ which generalizes Theorem 4.5.

(f) Changing the size of the cubes in defining the $\ell_1(L^2)$-norm yields equivalent norms. We will use this fact in proving Proposition 4.7. Notice also that translations on $\ell_1(L^2)$ act as uniformly bounded maps. This is *not* true of $L_\delta^2$ and is one reason for regarding $\ell_1(L^2)$ as being more natural than $L_\delta^2$.

PROOF OF THEOREM 4.1. We first prove that

$$\|f(x)g(-i\nabla)\|_2 = (2\pi)^{-\nu/2}\|f\|_2\|g\|_2 \tag{4.7}$$

For $f(x)g(-i\nabla)$ is an integral operator with kernel $(2\pi)^{-\nu/2} f(x)\check{g}(x-y)$, which has $L^2(\mathbb{R}^2)$ norm given by the right side of (4.7). Thus, (4.7) follows from Theorem 2.11. Clearly,

$$\|f(x)g(-i\nabla)\|_\infty \leq \|f\|_\infty\|g\|_\infty$$

It follows that if $f, g \in (L^\infty)_0$, the closure of $L^2 \cap L^\infty$ in $L^\infty$, then $f(x)g(-i\nabla) \in \mathcal{I}_\infty$. The general $p \geq 2$ case now follows by interpolation. Explicitly, we can find $a(x)$, $b(y)$ with $|a| = 1$, $|b| = 1$, so that $af = |f|$, $gb = |g|$. Thus, $|f|(x)|g|(-i\nabla)$ is $f(x)g(-i\nabla)$ multiplied by unitaries, so without loss, we can suppose that $f, g \geq 0$. Given $f, g \in L^\infty$ with compact support, map $0 \leq \operatorname{Re} z \leq \frac{1}{2}$ into $\mathcal{L}(\mathcal{H})$,

$$F(z) = f^{zp}(x)g^{zp}(-i\nabla)$$

Then $F(iy) \in \mathcal{I}_\infty$ with $\|F\|_\infty \leq 1$ and $F(\frac{1}{2}+iy) \in \mathcal{I}_2$ with $\|F(\frac{1}{2}+iy)\|_2 \leq (2\pi)^{-\nu/2}$. Complex interpolation (Theorem 2.9) now yields the result for such $f$ and $g$, and a limiting argument completes the proof. □

REMARK. The Hausdorff-Young inequality and Theorem 4.1 imply that if $A$ has integral kernel $f(x)h(x-y)$, then for $2<p<\infty$,

$$\|A\|_p \leq \|f\|_p\|h\|_p \tag{4.8}$$

Russo [**265, 267**] has generalized this result to general integral operators. Let $\|k\|_{p,q} = [\int[\int|k(x,y)|^p\,dx]^{q/p}\,dy]^{1/q}$. Then, if $A$ has kernel $k$ and $2\leq p<\infty$, we have

$$\|A\|_p \leq (\|k\|_{p',p}\|k^*\|_{p',p})^{1/2} \tag{4.9}$$

where $k^*(x,y)=\overline{k(y,x)}$. (4.9) implies (4.8) since $\|k\|_{p',p}=\|f\|_p\|h\|_p$, while $\|k^*\|_{p',p}$ is recognized as the $(p')$-th root of the $L^r$-norm of the convolution of $|f|^{p'}$ and $|h|^{p'}$, where $r=p/p'$, so $\|k^*\|_{p',p}\leq\|f\|_p\|h\|_{p'}$ by Young's inequality. (4.9) is proven by interpolating between the $p=2$ and $p=\infty$ results. Extensions and applications of these ideas can be found in papers of Russo and co-workers [**114, 171, 266, 268**]. Other results on general integral operators appear in Birman-Solomjak [**37, 38, 40, 44**].

PROOF OF THEOREM 4.5. We begin by proving Corollary 4.8 (following Kato [**166**]). Write $f(x)g(-i\nabla)=AB$ with

$$A=f(x)(1-\Delta)^{-\delta/2}(1+x^2)^{\delta/2} \qquad B=(1+x^2)^{-\delta/2}(1-\Delta)^{\delta/2}g(-i\nabla)$$

Since $g\in L^2_\delta$ and $\delta>\nu/2$, $B$ is of the form $F(x)G(-i\nabla)$ with $F,G\in L^2$ so $\|B\|_2\leq c\|g\|_{L^2_\delta}$, by Theorem 4.1. Let $h$ be $(2\pi)^{-\nu/2}$ times the Fourier transform of $(1+k^2)^{-\delta/2}$. Since $\widehat{h}$ is analytic in a tube with $\widehat{h}(\,\cdot\,+ia)\in L^2$, $h$ is in $L^2$ and $e^{a|x|}h(x)\in L^2$ for $a$ small. In particular, we have

$$\alpha\equiv\int(1+x^2)^\delta|h(x)|^2\,dx<\infty$$

$A$ is an integral operator with kernel $f(x)h(x-y)(1+y^2)^{\delta/2}$. Since $(1+y^2)^{\delta/2}\leq c(1+x^2)^{\delta/2}(1+(x-y)^2)^{\delta/2}$,

$$\int|f(x)h(x-y)(1+y^2)^{\delta/2}|^2\,dxdy\leq c\alpha\|f\|^2_{L^2_\delta}$$

so $\|A\|_2\leq d\|f\|_{L^2_\delta}$. Thus $AB\in\mathcal{J}_1$ and $\|AB\|_1\leq c\|f\|_{L^2_\delta}\|g\|_{L^2_\delta}$.

Now let $f_0,g_0$ be functions with support in $\Delta_0$. Then $\|f_0\|_{L^2_\delta}\leq c\|f_0\|_2$, etc., so for such $f_0$ and $g_0$, we have

$$\|f_0(x)g_0(-i\nabla)\|_1\leq c'\|f_0\|_2\|g_0\|_2$$

Let $f_\alpha(x)$ be supported in $\Delta_\alpha$, $g_\beta$ in $\Delta_\beta$. The unitary operator of translation in $x$ space by $\alpha$ units obeys $U_\alpha f_\alpha(x)U_\alpha^{-1}=\tilde{f}_\alpha(x)$ with $\operatorname{supp}\tilde{f}_\alpha\subset\Delta_0$, $U_\alpha g_\beta(-i\nabla)U_\alpha^{-1}=g_\beta(-i\nabla)$. If $V_\beta$ is translation in $p$ space, then

$$U_\alpha V_\beta(f_\alpha(x)g_\beta(-i\nabla))V_\beta^{-1}U_\alpha^{-1}=\tilde{f}_\alpha(x)\tilde{g}_\beta(-i\nabla)$$

so by the $\alpha=\beta=0$ result and the invariance of $\|\cdot\|_1$ under multiplication by unitaries,

$$\|f_\alpha(x)g_\beta(-i\nabla)\|_1\leq c'\|f_\alpha\|_2\|g_\beta\|_2$$

Given $f,g\in\ell^1(L^2)$, write $f_\alpha=\chi_\alpha f$, $g_\beta=\chi_\beta g$. Then

$$\|f(x)g(-i\nabla)\|_1\leq c'\sum_{\alpha,\beta}\|f_\alpha\|_2\|g_\beta\|_2=c'\|f\|_{2;1}\|g\|_{2;1}$$

This proves the theorem for $p = 1$. The result for $p = 2$ is given by (4.7) since $\ell^2(L^2) = L^2$. The general $1 < p < 2$ result follows by interpolation. $\square$

PROOF OF THEOREM 4.2 (CWIKEL [**80**]). Without loss, we can suppose that $f, g \geq 0$ as in the proof of Theorem 4.1. Also by homogeneity, we can suppose $\|f\|_p = \|g\|^*_{p,w} = 1$. For $n \in \mathbb{Z}$, define $f_n$ by

$$f_n(x) = \begin{cases} f(x) & \text{if } 2^{n-1} < f(x) \leq 2^n \\ 0 & \text{otherwise} \end{cases}$$

and similarly for $g_n$. For fixed $n \in Z$, let

$$A_n = \sum_{\ell+k\leq n} f_\ell(x) g_k(-i\nabla)$$

$$B_n = \sum_{\ell+k>n} f_\ell(x) g_k(-i\nabla)$$

(For the reader worried about the infinite sums here and below, we note that if $f$ and $g$ are simple functions, the sums are finite, and (4.5) for such functions implies (4.5) for all $f, g$ by a limiting argument.) Then

$$A_n + B_n = C \equiv f(x)g(-i\nabla)$$

Now for $\phi, \psi \in L^2$,

$$\begin{aligned} |(\phi, A_n\psi)| &\leq 2^n \sum_{\ell+k\leq n} \|2^{-\ell} f_\ell(x)\phi\| \, \|2^{-k} g_k(y)\widehat{\psi}\| \\ &\leq 2^n \Big(\sum_\ell \|2^{-\ell} f_\ell \phi\|^2\Big)^{1/2} \Big(\sum_k \|2^{-k} g_k \widehat{\psi}\|^2\Big)^{1/2} \\ &= 2^n \|\phi\| \, \|\psi\| \end{aligned}$$

where the last step follows from the fact that the $f_\ell$ (resp. $g_k$) have disjoint supports and $\sup_\ell \|2^{-\ell} f_\ell\|_\infty \leq 1$. Thus

$$\|A_n\|_\infty \leq 2^n$$

On the other hand,

$$\begin{aligned} \operatorname{Tr}(B_n^* B_n) &= \operatorname{Tr}\Big( \sum_{\substack{\ell+k>n \\ \ell'+k'>n}} f_\ell(x)\, \overline{f_{\ell'}(x)}\, g_k(-i\nabla)\, \overline{g_{k'}(-i\nabla)} \Big) \\ &= \operatorname{Tr}\Big( \sum_{\ell+k>n} |f_\ell(x)|^2 |g_k(-i\nabla)|^2 \Big) \\ &= \sum_{\ell+k>n} \|f_\ell(x) g_k(-i\nabla)\|_2^2 \\ &= (2\pi)^{-\nu} \sum_{\ell+k>n} \|f_\ell\|_2^2 \|g_k\|_2^2 \end{aligned}$$

(Use the disjointness of the supports in the second step and (4.7) in the last.) Now

$$\begin{aligned} \|g_k\|_2^2 &\leq 2^{2k}\, u\{x \mid g_k \geq 2^{k-1}\} \\ &\leq 2^{(2-p)k} 2^p \qquad (\text{since } \|g\|^*_{p,w} = 1) \end{aligned}$$

It follows that

$$\begin{aligned}
\|B_n\|_2^2 &\le c\sum_\ell \|f_\ell\|_2^2 \sum_{k=n+1-\ell} 2^{(2-p)k} \\
&\le c'\sum_\ell 2^{(2-p)(n-\ell)}\|f_\ell\|_2^2 && \text{(since } p>2) \\
&\le c''2^{(2-p)n}\sum_\ell \int (|f_\ell|^{p-2}|f_\ell|^2\,dx && \text{(since } f_\ell \ge 2^{\ell-1}) \\
&= c''2^{(2-p)n} && \text{(since } \|f_\ell\|_p = 1)
\end{aligned}$$

By Fan's inequality (Theorem 1.7), we have

$$\begin{aligned}
\mu_m(C) \le \|A_n\| + \mu_m(B_n) &\le \|A_n\|_\infty + m^{-1/2}\|B_n\|_2 \\
&\le 2^n + c''2^{(1-\frac{1}{2}p)n}m^{-1/2}
\end{aligned}$$

where we have used $\|B\|_2^2 \ge m\mu_m(B)^2$. Now, given $m$, choose $n$ so that $2^{n-1} \le m^{-1/p} < 2^n$. Then

$$\mu_m(C) \le (2+c'')m^{-1/p}$$

proving that $C \in \mathcal{I}_{p,w}$ as required. □

PROOF OF THEOREM 4.6. Without loss of generality, suppose that $\|f\|_{2;p} = 1$, $\|g\|^*_{2;p} = 1$. Write $f = \sum a_\alpha f_\alpha$ and $g = \sum b_\alpha g_\alpha$ where $\|f_\alpha\|_2 = \|g_\alpha\|_2 = 1$ and $f_\alpha, g_\alpha$ have support in $\Delta_\alpha, a_\alpha, b_\alpha \ge 0$. Then $a_\alpha \in \ell_p$ with $\ell_p$-norm one and $b_\alpha \in \ell_{p,w}$ with $\ell^*_{p,w}$-norm one. For each $n \in \mathbb{Z}$, let

$$\begin{aligned}
f_n &= \sum_{2^{n-1}<a_\alpha\le 2^n} a_\alpha f_\alpha \\
g_n &= \sum_{2^{n-1}<b_\alpha\le 2^n} b_\alpha g_\alpha \\
A_n &= \sum_{\ell+k\le n} f_\ell(x)g_k(-i\nabla) \\
B_n &= \sum_{\ell+k>n} f_\ell(x)g_k(-i\nabla)
\end{aligned}$$

Then, by the argument in the last proof,

$$\|A_n\|_2^2 = \sum_n a_\alpha^2 b_\beta^2$$

where the symbol $\sum_n$ indicates the sum over those $\alpha$ and $\beta$ with $a_\alpha \le 2^\ell$, $b_\beta \le 2^k$, and $\ell + k \le n$. The number of $b_\beta$ in $[2^{k-1}, 2^k]$ is bounded by $2^{-kp}$ so that

$$\begin{aligned}
\|A_n\|_2^2 &= \sum_\ell \Bigg[\sum_{2^{\ell-1}<a_\alpha\le 2^\ell} a_\alpha^2 \sum_{k=-\infty}^{n-\ell} 2^{(2-p)k}\Bigg] \\
&\le c\,2^{(2-p)n} \sum_{2^{\ell-1}<a_\alpha\le 2^\ell} a_\alpha^2 2^{-(2-p)\ell} && \text{(since } p<2) \\
&\le c'\,2^{(2-p)n}
\end{aligned}$$

By Theorem 4.5, we have

$$\|B_n\|_1 = \sum_n{}' a_\alpha b_\alpha$$

where $\sum_n'$ is the sum over $\alpha, \beta$ not summed in $\sum_n$. Proceeding as above, we have

$$\|B\|_1 \le c''\, 2^{(1-p)n}$$

By Fan's inequality, Theorem 1.7, we have

$$\mu_m(C) \le \mu_{m/2}(A_n) + \mu_{m/2}(B_n) \le d(m^{-1/2} 2^{(1-\frac{1}{2}p)n} + m^{-1} 2^{(1-p)n})$$

Choosing $2^{n-1} \le m^{1/p} \le 2^n$, we obtain $\mu_m(C) \le 3dm^{-1/p}$. □

Proof of Proposition 4.4. Let $f_n$ be defined by

$$f_n(x) = \begin{cases} f(x) & \text{if } |x| \text{ and } |f(x)| \le n \\ 0 & \text{otherwise} \end{cases}$$

and similarly for $g(y)$. Since $f_n, g_n \in L^2$, we have that $f_n(x)g_n(-i\nabla) \in \mathcal{J}_2$ and $\|f_n(x)g_n(-i\nabla)\| = (2\pi)^{-\nu/2}\|f_n\|_2\|g_n\|_2$ by (4.7). If $f(x)g(-i\nabla) \in \mathcal{J}_2$ also, then $\|f_n(x)g_n(-i\nabla)\|_2 \le \|f(x)g(-i\nabla)\|_2$ since $f_n g_n = A_n f g B_n$ for suitable contractions $A_n, B_n$. If $g \neq 0$, then $\|g_m\|_2 \neq 0$ for some $m$. Thus $\|f_n\|_2 \le (2\pi)^{\nu/2}\|g_m\|_2^{-1}\|f(x)g(-i\nabla)\|_2$, so $f \in L^2$. □

Proof of Proposition 4.7. By symmetry (since $-i\nabla$ goes into multiplication by $p$ under Fourier transform), it suffices to show that $f \in \ell^1(L^2)$ if $g \neq 0$. By the device above, we may suppose that $g \neq 0$ and that it is in $L^\infty$ with bounded support and, as usual, that $f, g \ge 0$. Then $h \equiv (2\pi)^{-\nu/2}\check{g}$ is positive definite and so we can pick a size for the basic cubes so that $\operatorname{Re}[h(x)] \ge a > 0$ for all $x$ in $\Delta_0$, or the $3^\nu - 1$ cubes surrounding $\Delta_0$. Now, by Proposition 2.6, if $g(-i\nabla)f(x)$ is trace class, we have

$$\sum_\alpha |\langle \chi_\alpha, h * (fp_\alpha)\rangle| < \infty$$

for any orthonormal set $p_\alpha$. Let $\tilde{\chi}_\alpha$ be the characteristic function of the union of $\Delta_\alpha$ and the $3^\nu - 1$ cubes surrounding it. Then, it follows that

$$\sum_\alpha |\operatorname{Re}\langle \tilde{\chi}_\alpha, h * (fp_\alpha)\rangle| < \infty$$

Thus, if each $p_\alpha$ is positive with support in $\Delta_\alpha$, we have

$$\sum_\alpha \|fp_\alpha\|_1 < \infty \tag{4.10}$$

since

$$\operatorname{Re}\langle \tilde{\chi}_\alpha, h * (fp_\alpha)\rangle = \int_{\substack{y\in\Delta_\alpha \\ x\in\tilde{\Delta}_\alpha}} (\operatorname{Re} h)(x-y) f(y) p_\alpha(y) \ge a \int (fp_\alpha)$$

Given $n$, let

$$f_{\alpha,n}(x) = \begin{cases} f(x) & \text{if } x \in \Delta_\alpha \text{ and } |f(x)| \le n \\ 0 & \text{otherwise} \end{cases}$$

Let $p_\alpha = \|f_{\alpha,n}\|_2^{-1} f_{\alpha,n}$ so that $\{p_\alpha\}$ is an orthonormal set; (4.10) then reads,

$$\sum_\alpha \|f_{\alpha,n}\|_2 < \infty$$

Taking $n \to \infty$, we find that $f \in \ell^1(L^2)$. □

This completes the proof of all the results stated at the start of this chapter. We close the chapter with a result about $f(x)g(-i\nabla)h(x)$:

THEOREM 4.9 (Birman-Solomjak [**40, 44**]). *If $g, f \in \ell^1(L^2)$ and $h \in \ell^\infty(L^2)$, then $f(x)g(-i\nabla)h(x) \in \mathcal{I}_1$ and*

$$\|f(x)g(-i\nabla)h(x)\|_1 \leq c\|f\|_{2;1}\|h\|_{2;\infty}\|g\|_{2;1}$$

REMARK. By interpolation and symmetry, we can take $g \in \ell^1(L^2)$, $f \in \ell^p(L^2)$, $h \in \ell^q(L^2)$ with $p^{-1} + q^{-1} = 1$.

PROOF. A result of Strichartz [**323**] proven below says that for $4m > \nu$, $\|(-\Delta+1)^{-m}h\|_{\text{op}} \leq c\|h\|_{2;\infty}$. Thus, if $f_0, g_0$ are in $L^2$ with support in $\Delta_0$, $\tilde{g}_0(y) = (1+y^2)^m g_0(y)$, then

$$\begin{aligned}\|f_0(x)g_0(-i\nabla)h(x)\|_1 &\leq \|f_0(x)\tilde{g}_0(-i\nabla)\|_1\|(-\Delta+1)^{-m}\|_{\text{op}} \\ &\leq d\|f_0\|_{2;1}\|g_0\|_{2;1}\|h\|_{2;\infty}\end{aligned}$$

Now we follow the translation argument used in proving Theorem 4.5. □

In the above, we used

LEMMA 4.10 (Strichartz [**323**]). *If $m$ is an integer with $4m > \nu$, then*

$$\|(-\Delta+1)^{-m}h\|_{\text{op}} \leq c\|h\|_{2;\infty}$$

PROOF. Since $(y^2+1)^{-m} \in L^2$, we have that $\|(-\Delta+1)^{-m}h\|_{\text{op}} \leq c\|h\|_2$ on account of Theorem 4.1 and $\|A\|_{\text{op}} \leq \|A\|_2$. Thus

$$\|h\phi\|_2 \leq c\|h\|_2\|(-\Delta+1)^m\phi\|_2 \tag{4.11}$$

Now let $\tilde{\Delta}_\alpha$ be the union of $\Delta_\alpha$ and its $3^\nu - 1$ neighbors. Pick $j_\alpha \in C_0^\infty$ so that $\operatorname{supp} j_\alpha \subset \tilde{\Delta}_\alpha$, $j_\alpha$ a translate of $j_0$, $0 \leq j_\alpha \leq 1$, and $\sum j_\alpha^2 = 1$. Let $\tilde{\chi}_\alpha$ be the characteristic function of $\tilde{\Delta}_\alpha$. Now, for any multi-index $\beta$ with $|\beta| < 2m$, we have $\|D^\beta\phi\|_2 \leq c_\beta\|(-\Delta+1)^m\phi\|_2$, so for any $\gamma$, $\sum_\alpha \|(D^\gamma j_\alpha)D^\beta\phi\|_2^2 \leq \|D^\gamma j_0\|_\infty^2 \sum_\alpha \|\tilde{\chi}_\alpha D^\beta\phi\|_2^2 \leq 3^\nu\|D^\gamma j_0\|_\infty^2\|D^\beta\phi\|_2^2 \leq c_{\beta,\gamma}\|(-\Delta+1)^m\phi\|^2$. It follows that

$$\sum_\alpha \|(-\Delta-1)^m j_\alpha\phi\|^2 \leq C\|(-\Delta+1)^m\phi\|_2^2$$

By (4.11), for any $h \in \ell^\infty(L^2)$,

$$\begin{aligned}\|h\phi\|_2^2 &= \sum_\alpha \|hj_\alpha\phi\|_2^2 \\ &\leq \sup_\alpha \|h\tilde{\chi}_\alpha\|_2^2 \sum_\alpha \|(-\Delta+1)^m j_\alpha\phi\|_2^2 \\ &\leq C3^\nu(\|h\|_{2,\infty})^2\|(-\Delta+1)^m\phi\|_2^2\end{aligned}$$

so that $\|(-\Delta+1)^{-m}h\|_{\text{op}} \leq 3^{\nu/2}C^{1/2}\|h\|_{2;\infty}$. (See Morgan [**225**] for improved constants.) □

In [**44**], Birman and Solomjak review and extend the theory of estimating singular values of integral operators. They include estimates that are "better than trace class," for example, $\sum(\mu_n(T))^p < \infty$ for $p < 1$ or $|\mu_n(T)| \leq Cn^{-\alpha}$, $\alpha > 1$, for suitable operators. See Addendum M for further discussion of "better than trace class" spaces.

CHAPTER 5

# Fredholm Theory

By "Fredholm theory" we mean explicit formulae for $(1+\mu A)^{-1} \equiv f(\mu)$ which work for all $\mu$ with $-\mu^{-1} \notin \sigma(A)$. $f(\mu)$ is not entire analytic for general $A$, but it is meromorphic, so one would like to write it as a ratio of entire analytic functions: $C(\mu)/B(\mu)$. $B(\mu)$ must have zeros where $f$ has poles, that is, at points where $(1+\mu A)$ does not have an inverse. This suggests that one take $B(\mu) = \det(1+\mu A)$. In fact, a detailed analysis [**164**] shows that $(1+\mu A)^{-1}$ has a pole at $-\mu_0^{-1} = \lambda_0 \in \sigma(A)$ of order not more than the algebraic multiplicity, so that $\det[1+\mu A)](1+\mu A)^{-1}$ is indeed an entire function. Rather than go through this analysis or follow the conventional analysis [**115, 209, 317**], we use a slightly different approach based on the fact that

$$\begin{aligned}\det(1+A+\lambda B) &= \det(1+A)\det(1+\lambda(1+A)^{-1}B) \\ &= \det(1+A)[1+\lambda\mathrm{Tr}((1+A)^{-1}B)+0(\lambda^2)] \qquad (5.1)\end{aligned}$$

which suggests that we obtain $(1+A)^{-1}\det(1+A)$ as a derivative of $\det(1+A)$. We begin with an abstract aside on analytic functions from one Banach space to another and their Fréchet derivatives:

DEFINITION. We call a function $f: X \to Y$, between Banach spaces, *finitely analytic* if and only if for all $A_1, \ldots, A_n \in X$, $f(\mu_1 A_1 + \cdots + \mu_n A)$ is an entire function (of $\mu_1, \ldots, \mu_n$) from $\mathbb{C}^n$ to $Y$.

DEFINITION. Let $f: X \to Y$ be a function between Banach spaces. Let $x_0 \in X$. We say that $T \in \mathcal{L}(X, Y)$ is the *Fréchet derivative* of $f$ at $x_0$ (and write $T = (Df)(x_0)$) if and only if

$$\|f(x+x_0) - f(x_0) - T(x)\| = o(\|x\|)$$

THEOREM 5.1. *Let $f$ be a finitely analytic function from $X$ to $Y$ obeying*

$$\|f(x)\| \le G(\|x\|) \qquad (5.2)$$

*for some monotone function $G$ on $[0,\infty)$. Then $f$ is Fréchet differentiable for all $x \in X$, $Df$ is a finitely analytic function from $X$ to $\mathcal{L}(X,Y)$, and*

$$\|(Df)(x)\| \le G(\|x\|+1) \qquad (5.3)$$

REMARK. From (5.3) and the fundamental theorem of calculus, we obtain

$$\|f(x)-f(y)\| \le \|x-y\| \int_0^1 G(\|tx+(1-t)y\|+1)\,dt \qquad (5.4)$$

which improves slightly the argument in the proof of Theorem 3.4. For example, using $\int_0^1 e^{at+(1-t)b}\,dt \le \max(e^a, e^b)$, one can improve (3.9) to read

$$|\det(1+A) - \det(1+B)| \le \|A-B\|_1 \exp(1+\max(\|A\|_1, \|B\|_1))$$

PROOF. Fix $x_0 \in X$. For each $x \in X$, let

$$T(x) = \left. \frac{df(x_0 + \mu x)}{d\mu} \right|_{\mu=0}$$

Then $x \to T(x)$ is easily seen to be linear in $x$ and moreover, by a Cauchy estimate, for $\|x\| = 1$,

$$\|T(x)\| \leq \sup_{|\mu|=1} \|f(x_0 + \mu x)\| \leq G(\|x_0\| + 1)$$

Finally, by a second Cauchy estimate for $|\mu| < 1$, $\|x\| = 1$,

$$\|f(x_0 + \mu x) - f(x_0) - \mu T(x)\| \leq G(\|x\| + 1)(2\pi)^{-1} \int_0^{2\pi} d\theta\, g_\mu(\theta)$$

where $g_\mu(\theta) = |e^{-i\theta} - (e^{i\theta} - \mu)^{-1} + \mu e^{-2i\theta}| = 0(|\mu|^2)$, proving that $T$ is the Fréchet derivative. □

From this theorem, (3.6), and (5.1), we immediately conclude:

COROLLARY 5.2. *$f : A \mapsto \det(1 + A)$ as a function from $\mathfrak{I}_1$ to $\mathbb{C}$ is Fréchet differentiable and, under the identification of $\mathfrak{I}_1^*$ and $\mathcal{L}(\mathcal{H})$, its derivative is given by*

$$Df(A) = (1 + A)^{-1} \det(1 + A)$$

*for those $A$ with $-1 \notin \sigma(A)$. In particular, the function*

$$D(A) \equiv -A(1 + A)^{-1} \det(1 + A) \tag{5.5}$$

*(the first Fredholm minor) defines a finitely analytic function from $\mathfrak{I}_1$ to $\mathfrak{I}_1$ obeying*

$$\|D(A)\|_1 \leq \|A\|_1 \exp(\|A\|_1 + 1) \tag{5.6}$$

*and*

$$\|D(A) - D(B)\|_1 \leq \|A - B\|_1 \exp(2 + \max(\|A\|_1, \|B\|_1)) \tag{5.7}$$

The point of this definition is that

$$(1 + \mu A)^{-1} = 1 + \frac{\mu D(\mu A)}{\det(1 + \mu A)} \tag{5.8}$$

REMARKS. 1. By more detailed estimates [**288**], the $\exp(\|A\|_1 + 1)$ in (5.6) can be replaced by $\exp(\|A\|_1)$ and then the 2 in (5.7) by 1.

2. One can also define the higher Fredholm minors, which are related to $\Lambda^k([1+A]^{-1}) \det(1 + A)$, by this method. By Theorem 5.1, any $f$ of the given form has Fréchet derivatives of all orders. The $k$-th derivative is a map in $\mathcal{L}_k(X, Y)$ defined inductively by $\mathcal{L}_1(X, Y) = \mathcal{L}(X, Y)$ and $\mathcal{L}_k(X, Y) = \mathcal{L}(X, \mathcal{L}_{k-1}(X, Y))$. It is not hard to see that $\mathcal{L}_k(X, Y)$ can be associated with $k$-linear maps from $X$ to $Y$. Thus $(D^k f)(x_0)$ is a $k$-linear map of norm bounded by $g(\|x_0\| + k)$. (If $g(y) \leq e^{ay}$, then the Taylor series for $f$ is automatically convergent.) One can easily show, by extending (5.1) to higher order, that

$$(D^k f)(A_0) : B_1, \ldots, B_k \mapsto k!\, \mathrm{Tr}((B_1 \otimes \cdots \otimes B_k)\Lambda^k((1 + A)^{-1})) \det(1 + A)$$

when $f(A) = \det(1 + A)$ as above.

The point of the estimates (5.6) on $D(\mu A)$ and (3.6) on $\det(1+\mu A)$ is that they allow one to control the rate of convergence of $D(\mu A)$ and $\det(1+\mu A)$, and obtain explicit estimates on the errors obtained by truncating Taylor series. It is obviously of some interest to obtain useful expressions for Taylor coefficients of $D(\mu A)$ and $\det(1+\mu A)$. Here we consider first general formulae obtained by Plemelj [**240**] and rediscovered by Smithies [**316**], and then supplement the Fredholm formula (3.14) for $\det(1+\mu A)$ when $A$ is an integral operator by a similar formula for $D(\mu A)$.

LEMMA 5.3. *Let $f$ and $g$ be functions analytic near $z$ with Taylor expansions $f(z) = \sum_{m=0}^{\infty} a_n z^n/n!$, $g(z) = \sum_{n=1}^{\infty} (-1)^{n+1} b_n z^n/n$. Suppose that $f(z) = \exp(g(z))$. Then $a_n$ is the $n \times n$ determinant,*

$$a_n = \begin{vmatrix} b_1 & n-1 & 0 & \dots & 0 \\ b_2 & b_1 & n-2 & \dots & 0 \\ \vdots & \ddots & \ddots & \ddots & \vdots \\ b_{n-1} & b_{n-2} & b_{n-3} & \dots & 1 \\ b_n & b_{n-1} & b_{n-2} & \dots & b_1 \end{vmatrix} \tag{5.9}$$

PROOF. Clearly, $a_0 = 1$ and since $f'(z) = f(z)g'(z)$, the $a$'s obey

$$a_n = \sum_{k=1}^{n} b_k a_{n-k} (-1)^{k+1} \frac{(n-1)!}{(n-k)!} \tag{5.10}$$

(5.10) determines the $a$'s inductively. Expanding the determinant (5.9) in minors in the first column, we see that the right side of (5.9) obeys (5.10). □

THEOREM 5.4 (Plemelj-Smithies Formulae). *Define $\alpha_m(A)$, $\beta_m(A)$ for $A \in \mathcal{J}_1$ by $\det(1+\mu A) = \sum_{m=0}^{\infty} \mu^m \alpha_m(A)/m!$ and $D(\mu A) = \sum_{m=0}^{\infty} \mu^{m+1} \beta_m(A)/m!$. Then*

$$\alpha_m(A) = \begin{vmatrix} \operatorname{Tr}(A) & m-1 & \dots & 0 \\ \operatorname{Tr}(A^2) & \operatorname{Tr}(A) & \dots & 0 \\ \vdots & \ddots & \ddots & \vdots \\ \operatorname{Tr}(A^m) & \operatorname{Tr}(A^{m-1}) & \dots & \operatorname{Tr}(A) \end{vmatrix}_{m\times m}$$

$$\beta_m(A) = \begin{vmatrix} A & m & 0 & \dots \\ A^2 & \operatorname{Tr}(A) & m-1 & \dots \\ \vdots & \vdots & \ddots & \ddots \\ A^{m+1} & \operatorname{Tr}(A^m) & \dots & \operatorname{Tr}(A^{m-1}) \end{vmatrix}_{(m+1)\times(m+1)}$$

PROOF. Clearly, for small $\mu$,

$$D(\mu A) = (\mu A - \mu^2 A^2 + \mu^3 A^3 \dots) \det(1+\mu A)$$

so

$$\beta_m(A) = m! \left[ A \frac{\alpha_m(A)}{m!} - A^2 \frac{\alpha_{m-1}(A)}{(m-1)!} + \cdots \right] \tag{5.11}$$

Expanding the determinant for $\beta_n$ in minors about the first column, we see that it is sufficient to prove the formula for $\alpha_m(A)$. This formula follows from the lemma if we show that for $\mu$ small,

$$\det(1+\mu A) = \exp\left( \sum_{m=1}^{\infty} \frac{(-1)^{m+1}}{m} \mu^m \operatorname{Tr}(A^m) \right) \tag{5.12}$$

Now, for $|\mu|\,\|A\|_1 < 1$,

$$\sum_{m=1}^{\infty} \frac{(-1)^{m+1}}{m} \mu^m \mathrm{Tr}(A^m) = \sum_{m=1}^{\infty} (-1)^{m+1} \mu^m \sum_n \frac{1}{m} \lambda_n(A)^m$$
$$= \sum_n \log(1 + \mu\lambda_n(A))$$

where we use Lidskii's equality, (3.2), in the first step and the inequality $\sum |\lambda_n(A)|^m \leq \|A^m\|_1 \leq \|A\|_1^m$ to interchange the sums. Thus the right side of (5.12) is $\prod_m (1 + \mu\lambda_m(A)) = \det(1 + \mu A)$ by (3.11). □

THEOREM 5.5 (Fredholm's Formula). *Let $A$ in $\mathcal{I}_1$ have the form of (3.12) and let $K\left(\begin{smallmatrix} x_1 & \cdots & x_n \\ y_1 & \cdots & y_n \end{smallmatrix}\right)$ be given by (3.14). Then $D(\mu A) = \sum_{n=0}^{\infty} \mu^{m+1} \beta_m(A)/m!$ where $\beta_m(A)$ is an integral operator with kernel $K_m(x, y)$ given by*

$$K_m(x, y) = \int_a^b dx_1 \ldots \int_a^b dx_m \, K\begin{pmatrix} x & x_1 & \ldots & x_m \\ y & x_1 & \ldots & x_m \end{pmatrix} \tag{5.13}$$

PROOF. We expand $K$ as follows: Any permutation $\pi$ on $1, \ldots, m+1$ can be written $1 \to i_1 \to \cdots \to i_{k-1} \to 1$ times a permutation on $n + 1 - k$ objects. Thus the right side of (5.13) is

$$\sum_{k=1}^{m} \int_a^b \cdots \int_a^b (-1)^{k-1} c_{mk} \, K(x, x_1) \ldots K(x_{k-1}, y) K\begin{pmatrix} x_k & \ldots & x_m \\ x_k & \ldots & x_m \end{pmatrix}$$

where $c_{mk}$ is the number of ways of choosing $i_1, \ldots, i_{k-1}$, that is, $m(m-1)\ldots(m-k+1)$. Using (5.11) and (3.14), we see that this gives $K_m(x, y)$. □

REMARK. (5.13) has an abstract version in terms of partial traces of $\Lambda^{m+1}(A)$; see [**300**].

As an application of this machinery, we consider the Lippmann-Schwinger equation [**207**] of scattering theory. We will confine our discussion to the equations themselves. See [**253, 135, 229**] for the connection with scattering theory.

Consider $L^2(\mathbb{R}^\nu)$. For $\alpha$ with $\mathrm{Im}\,\alpha > 0$, $(-\Delta - \alpha^2)^{-1}$ is an integral operator with kernel $G_0(x, y; \alpha)$; for example, if $\nu = 1$ or 3,

$$(\nu = 1) \qquad G_0(x, y; \alpha) = (-2i\alpha)^{-1} \exp(i\alpha|x - y|) \tag{5.14}$$

$$(\nu = 3) \qquad G_0(x, y; \alpha) = (4\pi|x - y|^{-1}) \exp(i\alpha|x - y|) \tag{5.15}$$

For $\nu \neq 1$ or 3, $G_0$ is a Bessel function but has the same qualitative features; explicitly, it only depends on $|x-y|$, falls at infinity more or less like $\exp(-(\mathrm{Im}\,\alpha)|x-y|)$, and has a singularity at $x - y = 0$ like $|x - y|^{-(\nu-2)}$ $(\nu \geq 3)$ or $\log(|x - y|^{-1})$ $(\nu = 2)$. Moreover, $G_0$ has a limit as $\alpha$ goes to a non-zero real and if $\nu \geq 3$, as $\alpha$ goes to zero. We continue to denote this by $G_0(x, y; \alpha)$. For $\nu \geq 3$, the kernel, $G_0(x, y; \alpha = 0)$, of $(-\Delta)^{-1}$ has the form

$$G_0(x, y; \alpha = 0) = c_\nu |x - y|^{-(\nu-2)} \tag{5.16}$$

(5.16) can be proven by noting that $G_0(\,\cdot\,,\,\cdot\,; \alpha = 0)$ is, up to a constant, $\check{h}(x - y)$ where $h(k) = k^2$; $\check{h}$ can be computed easily by noting it is rotation invariant and scales in a certain way.

For $k \in \mathbb{R}^\nu$, let $\phi_0(x,k) = \exp(ik \cdot x)$. Then the Lippmann-Schwinger equation reads:

$$\phi(x;k) = \phi_0(x;k) - \int G_0(x,y;|k|)V(y)\phi(y;k)\,dy \tag{5.17a}$$

or formally (putting $H_0 = -\Delta$),

$$\phi = \phi_0 - (H_0 - |k|^2)^{-1} V\phi \tag{5.17b}$$

(5.17b) implies that formally $(H_0 - |k|^2)\phi = -V\phi$, so that $\phi$ should be a generalized eigenfunction for $H = H_0 + V$ and could play the same role in an eigenfunction transform that $\phi$ plays in the Fourier transform. Moreover, $\phi$ determines the $T$-matrix,

$$T(k,k') = (2\pi)^{-\nu} \int \overline{\phi_0(x,k)}\, V(x)\phi(x,k')\,dx \tag{5.18}$$

which is related to the scattering operator.

Since $\phi_0 \notin L^2$, we cannot use the simplest kind of Hilbert space methods on (5.17) directly so, following Grossman-Wu [**141, 142**], we introduce $\psi_0(x,k) = |V(x)|^{1/2}\phi_0(x,k)$; $\psi(x,k) = |V(x)|^{1/2}\phi(x,k)$; $V^{1/2} = V/|V|^{1/2}$, so that (5.7) becomes

$$\psi(x,k) = \psi_0(x,k) - \int |V(x)|^{1/2} G_0(x,y;|k|)V^{1/2}(y)\psi(y,k)\,dy \tag{5.19a}$$

If $V \in L^1$, then $\psi_0 \in L^2$, so we can hope to solve (5.18) by

$$\psi = (1+K)^{-1}\psi_0 \tag{5.19b}$$

where $K$ has integral kernel $|V|^{1/2}G_0V^{1/2}$. For clarity, we consider the one-parameter family of potentials $\lambda V$ so that (5.19b) becomes $\psi = (1+\lambda K)^{-1}\psi_0$. Then (5.18) has the formal expansion

$$(2\pi)^\nu T = \sum_{n=0}^{\infty} (-\lambda)^n (\psi_0, K^n \psi_0) \tag{5.20}$$

This series is known as the *Born expansion* for $T$. Under suitable assumptions, this expansion can be shown to converge for $\lambda$ sufficiently small [**277**] or $|k|$ sufficiently large [**344**] (in the large $k$ region, $G_0$ oscillates so a Riemann–Lebesgue-type phenomenon takes over). The point is that, even if $(1+K)$ is invertible, the series (5.20) may diverge (and will) if $\sigma_r(K)$, the spectral radius of $K$, is larger than 1. This is precisely the situation where Fredholm theory is useful.

Consider first the case $\nu = 1$:

PROPOSITION 5.6. *Let $\int (1+|x|^{1+\delta})|V(x)|\,dx < \infty$ for some $\delta > 0$. Then, for any real $k \neq 0$, the kernel*

$$K(x,y;k) = |V(x)|^{1/2} \exp(ik|x-y|)V^{1/2}(y)$$

*defines a trace class operator on $L^2(\mathbb{R}, dx)$.*

PROOF. Write $K = \sum K_{\alpha\beta}$ where $K_{\alpha\beta}$ is the kernel restricted to $\alpha < x \leq \alpha+1$, $\beta < y \leq \beta+1$. For $\alpha < \beta$, $|x-y| = y-x$, so $K_{\alpha\beta}$ is rank one. It is thus clearly trace class and its trace class norm is $a_\alpha a_\beta$ where $a_\alpha = (2|k|)^{-1/2}(\int_\alpha^{\alpha+1} |V(x)|\,dx)^{1/2}$. A similar result holds for $\beta < \alpha$, so we need only consider the case $\alpha = \beta$. Consider the function $h(z) = (-2ik)\exp(ik|z|)$ for $|z| \leq 2$. Extend $h$ to be $C^\infty$ outside of $z = 0$ and to have compact support. Then $h \in L^2$ and its distributional derivative is in $L^2$, so $h = (2\pi)^{-1/2}\check{g}$ for a $g(k)$ obeying $g \in L^2$, $kg \in L^2$; thus $g \in L^1$.

It follows that $K_{\alpha\alpha} = |V_\alpha(x)|^{1/2} g(-i\nabla) V_\alpha^{1/2}(x)$ where $V_\alpha$ is the obvious function. By Theorem 4.1, $K_{\alpha\alpha}$ is trace class with norm bounded by $ca_\alpha^2$, Thus $\|K\|_1 \leq c\sum_{\alpha,\beta} a_\alpha a_\beta = c(\sum a_\alpha)^2$. Since $\int_a^{\alpha+1} |V(x)|\, dx \leq c(1+|\alpha|)^{-1-\delta} \int_\alpha^{\alpha+1} |V(x)|(1+|x|^{1+\delta}\, dt$, we have that $\sum a_\alpha < \infty$, as in remark (c) following Proposition 4.7. □

REMARK. Formally, $K$ is of the form $h(x)g(-i\nabla)f(y)$ were $g$ is not a function but rather a distribution, so this result in some sense extends to those of Chapter 4.

As a result, we obtain a different expression for $T$:

$$(2\pi)^\nu T = (\psi_0, \psi_0) + [\det(1+K)]^{-1}(\psi_0, D(K)\psi_0) \tag{5.21}$$

Both $\det(1+K)$ and $(\psi_0, D(K)\psi_0)$ are given by series which converge for all $k$ and $\lambda$ so that (5.21) is more satisfactory than (5.20). Moreover, explicit error estimates are possible.

For $\nu > 1$, the kernel $K$ will not be trace class (formally $\mathrm{Tr}(K) = \int V(x) G_0(0,0;|k|)\, dx$ and $G_0(0,0;|k|) = \infty$ if $\nu > 1$). But for $\nu = 3$, $K$ will be Hilbert-Schmidt so long as

$$\int \frac{|V(x)|\,|V(y)|}{|x-y|^2}\, dxdy < \infty$$

and this will hold if $V \in L^{3/2}$ (by a Sobolev inequality). For general $\nu$, $K$ will be in some $\mathcal{I}_p$ space. This suggests that it would be useful to extend Fredholm theory from $\mathcal{I}_1$ to $\mathcal{I}_p$. Historically, there were two approaches to this problem:

POINCARÉ'S METHOD. In a paper which has largely been ignored, Poincaré [**241**] remarked that if $-A^2$ were an operator to which Fredholm theory applied, then one could try to find $(1+A)^{-1}$ by using

$$(1+A)^{-1} = (1-A)(1-A^2)^{-1} \tag{5.22}$$

Thus one has a method for finding $(1+A)^{-1}$ if $A \in \mathcal{I}_2$ (of course, Poincaré did not know about $\mathcal{I}_1$ or $\mathcal{I}_2$). The biggest advantage of Poincaré's method is that no new estimates are needed. The main disadvantage is that if $(1-A)$ fails to be invertible, the right side of (5.22) won't make sense. A secondary disadvantage is that the formulae are somewhat complicated.

HILBERT'S METHOD. An approach that Hilbert [**149**] and later Carleman [**68**] and Smithies [**316, 317**] developed for $\mathcal{I}_2$ has been extended to $\mathcal{I}_p$ [**55, 105, 134**]. This approach begins by remarking that in the small $\mu$ expansion, $\det(1+\mu A) = \exp(\mu \mathrm{Tr}(A) - \frac{\mu^2}{2}\mathrm{Tr}(A^2) + \cdots)$, only $\mathrm{Tr}(A)$ is singular for $A \in \mathcal{I}_2$. Thus we might define

$$\det{}_2(1+A) = \det(1+A)e^{-\mathrm{Tr}(A)} \tag{5.23}$$

originally on $\mathcal{I}_1$. While neither $\det(1+A)$ nor $\mathrm{Tr}(A)$ can be extended to $\mathcal{I}_2$, one might hope to extend $\det_2$ to $\mathcal{I}_2$ by developing suitable estimates; this can be done. The main advantages of this program is that it turns out that $\det_2(1+A) = 0$ if and only if $(1+A)$ fails to have an inverse and the formulae are simple. Its main disadvantage is that new estimates are required in its conventional treatment.

In Chapter 9, we present a method originally developed by Seiler [**285, 288**], which is a synthesis of the Poincaré and Hilbert approaches. Rather than deal with $(1-A^2)$ à la Poincaré, we consider $B = (1+A)e^{-A}$. Clearly, $B$ is invertible if and only if $(1+A)$ is invertible so that the spurious zeros problem of Poincaré is absent.

Moreover, $\tilde{A} = B - 1$ obeys $\tilde{A} = A^2 g(A)$ for a suitable entire function $g$. Since $A^2 \in \mathcal{J}_1$, we have $\tilde{A} \in \mathcal{J}_1$ so $\det(1+\tilde{A})$ can be formed without any new estimates. Of course, if $A \in \mathcal{J}_1$, we have

$$\det(1+\tilde{A}) = \det(1+A)\det(e^{-A}) = \det(1+A)e^{-\operatorname{Tr}(A)}$$

since $\det(e^{-A}) = \prod_n (1+\lambda_n(e^{-A}-1)) = \prod_n \lambda_n(e^{-A}) = e^{-\sum \lambda_n(A)} = e^{-\operatorname{Tr}(A)}$. Thus $\det(1+\tilde{A})$ is precisely the $\det_2$ of the Hilbert program.

This $\det_2$ can be used to find a solution of the $\nu = 3$ Lippmann-Schwinger equation and the more general $\det_n$ of Chapter 9 for the higher $\nu$ Lippmann-Schwinger equation. These relations of the $T$-matrix as a rational function suggest that Padé approximants built from the Born series might have more luck than the Born series in recovering $T$; this is sometimes the case [**71, 121**].

We remark that the Grossman-Wu factorization is not absolutely necessary. By using Fredholm theory on a Banach space, Ikebe [**153**] solved the $\nu = 3$ Lippmann-Schwinger equation directly and developed the associated eigenfunction transform before the corresponding factorized approach was developed in [**292**]. Moreover, Agmon's elegant approach [**3**] can be viewed as the solution of the Lippmann-Schwinger equation on the Hilbert space $L^2_\delta$ (with the additional trick that $k$ is averaged over angles).

Finally, we want to describe a connection between the Fredholm determinant in the $\nu = 1$ case and the Wronskian of suitable solutions on the one-dimensional Schrödinger equation. If one worked on the half-line, one would realize the famous Jost function [**84, 135, 229, 253**] as a Fredholm determinant. Similarly, the discriminant of Floquet theory [**107, 254**] is a Fredholm determinant.

Fix $k$ with $\operatorname{Im} k \geq 0$, $k \neq 0$, and $\lambda$ in $\mathbb{C}$. For regular enough $V$'s, the integral equation,

$$\phi(x) = e^{ikx} - \lambda \int_x^\infty k^{-1} \sin k(x-y) V(y)\phi(y)\, dy$$

is easily shown to have a solution by just summing the iterates: The $n$-th term is bounded by $[c(x,k)]^n |\lambda|^n / n!$ for suitable $c(x,k)$; see [**135, 253**]. The corresponding solution $\phi_+(x;k,\lambda)$ thus obeys

$$|\phi_+(x;k,\lambda| \leq \exp(|\lambda| c(x,k))$$

and $\phi_+$ is easily seen to obey the Schrödinger equation $-u'' + \lambda V u = k^2 u$. Similarly, one can find a solution $\phi_-(x;k,\lambda)$ which looks like $e^{-ikx}$ as $x \to -\infty$. Define

$$D(k,\lambda) = (2ik)^{-1}[\phi_+'(x)\phi_-(x) - \phi_+(x)\phi_-'(x)]$$

the *Wronskian* of $\phi_+$ and $\phi_-$; this is independent of $x$. The point is:

PROPOSITION 5.7. *$D(k,\lambda)$ is the determinant of $1 + \lambda K(\,\cdot\,,k)$ with $K(\,\cdot\,,k)$ given by Proposition 5.6.*

PROOF. Let $\widetilde{D}(k,\lambda) = \det(1+\lambda K(\,\cdot\,,k))$. Both $D$ and $\widetilde{D}$ can be shown to be analytic in $\operatorname{Im} k > 0$ and continuous in $\operatorname{Im} k \geq 0$, $k \neq 0$, so we need only prove the result for fixed $k$ purely imaginary. Now $\widetilde{D}(k,\lambda) = 0$ if and only if $-\lambda K\phi = \phi$ has a solution and this, as we shall see in Chapter 7, happens only if $-u'' + \lambda V u = k^2 u$ has a square integrable solution. But for $\operatorname{Im} k > 0$, there is only one solution square integrable at $+\infty$, namely, $\phi_+$, and only one at $-\infty$, namely, $\phi_-$, so $-u'' + \lambda V u = k^2 u$ has a square integrable solution if and only if $D(k,\lambda) = 0$. Thus $D$ and $\widetilde{D}$ have the same zeros for $k$ fixed (and a further analysis shows these zeros to be simple). Since both $D$ and $\widetilde{D}$ are exponentially bounded, it follows that for fixed $k$,

$$D(\lambda) = \widetilde{D}(\lambda)e^{a+b\lambda}$$

But, by normalization of $D$, $D(0) = 1$ and $\widetilde{D}(0) = 1$, so $a = 0$. Moreover, by an explicit calculation, both $D$ and $\widetilde{D}$ begin

$$1 + \lambda(-2ik)^{-1} \int_{-\infty}^{\infty} V(x)\, dx + 0(\lambda^2)$$

so that $b = 0$. □

CHAPTER 6

# Scattering With a Trace Condition

There is an enormous literature on mathematical scattering theory — it is possible to fill a whole book on the subject [**253**] and really only scratch the surface. Obviously, I can only give the barest taste of the subject in a single lecture. I want to concentrate on a subset of the subject initiated by Kato and Rosenblum and systematically developed by a variety of people, most notably Birman, Kato, and Kuroda; complete references can be found in [**253**].

In many problems in physics, the dynamics is described by a map, $\phi \mapsto \phi_t$, on the data which conserves a positive quadratic functional. This is true in quantum theory where the functional is $\int |\phi|^2\, dx$ but also for classical linear wave equations; for example, if $\ddot{u} = \Delta u$ and $\phi = \langle u, \dot{u}\rangle$, then the energy $\int (|\dot{u}(x)|^2 + |\nabla u|^2\, dx$ is conserved. In this situation, one uses the functional to form a Hilbert space and the dynamics is given by a unitary group $\phi_t = e^{-iAt}\phi$ for some (unbounded) self-adjoint operator $A$. Recall [**250**] that, for each $\phi$, there is (either by the spectral theorem or by Bochner's theorem) a measure $d\nu_\phi$ so that

$$(\phi, e^{-iAt}\phi) = \int e^{-ixt}\, d\nu_\phi(x) \tag{6.1}$$

The set for all $\phi$ for which $d\nu_\phi$ is absolutely continuous with respect to Lebesgue measure is a closed subspace $\mathcal{H}_{\text{ac}}$, and we use $P_{\text{ac}}(A)$ to stand for the orthogonal projection onto $\mathcal{H}_{\text{ac}}(A)$.

Typically, solutions of "interacting" dynamics $e^{-iAt}\phi$ look asymptotically like a simpler "free" dynamics $e^{-iBt}\psi_\pm$ as $t \to \mp\infty$. Since $\|e^{-iAt}\phi - e^{-iBt}\psi_+\| \to 0$ as $t \to -\infty$ if and only if $\|\phi - e^{+iAt}e^{-iBt}\psi_+\| \to 0$, scattering theory depends on the existence of the limits

$$\Omega^\pm(A,B) = \operatorname*{s-lim}_{t\to\mp\infty} e^{iAt}e^{-Bt}P_{\text{ac}}(B) \tag{6.2}$$

The funny $\mp\infty$ convention is used to obtain agreement with the physics literature. (Unfortunately, this results in the disagreement with the convention of most mathematicians, but that is okay since mathematicians are more flexible than physicists.) The projection $P_{\text{ac}}(B)$ is a natural object to put in, since the point spectrum is unstable under perturbation (e.g., if $B\phi = \lambda\phi$, then $e^{iAt}e^{-iBt}\phi$ has a limit if and only if $A\phi = \lambda\phi$) and the singular continuous spectrum is also unstable (see, e.g., [**102**] and Section 12.2). Finally, we remark that for the classical linear wave equations alluded to above, the free and interacting energies are different so that $e^{-iAt}$ and $e^{-iBt}$ act on different Hilbert spaces. One must then modify (6.2) and use a theory pioneered by Birman, Kato, and Wilcox. Trace class ideas are applicable in this situation and are admirably reviewed in the monograph of Deift [**86**]. Lax and Phillips have developed an alternative approach to scattering of these classical equations (see [**189**] or [**253**, Section XI.11]). In applications below, we will only consider the single Hilbert space theory.

An important goal of the theory is to prove that $\operatorname{Ran}\Omega^{\pm}(A,B) = \mathcal{H}_{\text{ac}}(A)$, in which case we say that $\Omega^{\pm}(A,B)$ are *complete*. Completeness implies that $A \restriction \mathcal{H}_{\text{ac}}(A)$ and $B \restriction \mathcal{H}_{\text{ac}}(B)$ are unitarily equivalent operators. Moreover, it says that any $\phi$ obeying $e^{-iAt}\phi \sim e^{-iBt}\psi_+$ as $t \to -\infty$ for some $\psi_+ \in \mathcal{H}_{\text{ac}}(B)$ also obeys $e^{-iAt}\phi \sim e^{-iBt}\psi_-$ as $t \to +\infty$; that is, any interacting state which is asymptotically free in the past, is free in the future — obviously an important factor for the physical interpretation. Finally, in case $\mathcal{H}_{\text{ac}}(B) = \mathcal{H}$, completeness implies that the $S$-matrix, $S = \Omega^-(A,B)^*\Omega^+(A,B)$, is a unitary operator.

The following result is very simple but is remarkable in that it reduces the completeness question to an existence question.

PROPOSITION 6.1. *Suppose that $\Omega^{\pm}(A,B)$ exist. Then they are complete if and only if $\Omega^{\pm}(B,A)$ exist.*

PROOF. Let $\psi \in \mathcal{H}_{\text{ac}}(B)$. Then $\lim_{t\to-\infty} e^{iAt}e^{-iBt}\psi$ exists and equals $\phi$ if and only if $\|e^{-iAt}\phi - e^{-iBt}\psi\| \to 0$ if and only if $\psi = \lim_{t\to-\infty} e^{iBt}e^{-iAt}\phi$. Since $\phi$ must be in $\mathcal{H}_{\text{ac}}(A)$ [this follows from the calculation $(\phi, e^{-iAs}\phi) = \lim_{t\to\infty}(e^{iA(t-s)}e^{-iB(t-s)}\psi, e^{-iAs}e^{iAt}e^{-iBt}\psi) = (\psi, e^{-iBs}\psi)$], we see that for $\psi \in \mathcal{H}_{\text{ac}}(B)$, $\lim_{t\to-\infty} e^{iAt}e^{-iBt}\psi$ exists if and only if $\psi \in \operatorname{Ran}\Omega^+(A,B)$. Since the range is automatically in $\mathcal{H}_{\text{ac}}(B)$ (by the above calculation in [...]), the proposition follows. $\square$

The heart of the trace class scattering theory is:

THEOREM 6.2. *Let $A$ and $B$ be self-adjoint operators and $J$ a bounded operator so that*

$$AJ - JB \in \mathcal{J}_1 \tag{6.3}$$

*Then*

$$\Omega^{\pm}(A,B;J) \equiv \operatorname*{s-lim}_{t\to\mp\infty} e^{iAt}Je^{-iBt}P_{\text{ac}}(B)$$

*exists.*

REMARKS. 1. This version of the Kato-Birman theory was emphasized by Pearson [**232**], who is also responsible for the proof we give below.

2. (6.3) is intended in the sense that there is a $C \in \mathcal{J}_1$ so that for any $\phi \in D(B)$ and $\psi \in D(A)$, $(\psi, C\phi) = (A\psi, J\phi) - (J\psi, B\phi)$. Henceforth, we will ignore all domain subtleties.

The proof of Theorem 6.2 will depend on two lemmas, each of which goes back to the earliest days of the Kato-Birman theory:

LEMMA 6.3. *Let $C$ be compact and $A$ self-adjoint. Then*

$$\operatorname*{s-lim}_{t\to\pm\infty} Ce^{-iAt}P_{\text{ac}}(A) = 0 \tag{6.4}$$

*This holds more generally if $C$ is bounded and $C(A+i)^{-n}$ is compact for some $n$.*

PROOF. By a limiting argument, it suffices to prove (6.4) when $C$ is finite rank and so when $C$ is rank one. Thus it suffices to show that $\operatorname{w-lim} e^{-iAt}P_{\text{ac}}(A) = 0$. But if $\phi \in \mathcal{H}_{\text{ac}}(A)$, $(\psi, e^{-iAt}\phi) = \int e^{-ixt}\,d\mu_{\psi,\phi}$ with $d\mu_{\psi,\phi}(x) = F(x)\,dx$ and $\int |F(x)|\,dx < \infty$ (see below), so that $(\psi, e^{-iAt}\phi) \to 0$ by the Riemann-Lebesgue lemma.

Given the result for compact $C$, suppose that $C' = C(A+i)^{-n}$ is compact. Then, by (6.4) for $C'$, $Ce^{-iAt}P_{\text{ac}}(A)\phi = C'e^{-iAt}P_{\text{ac}}(A)(A+i)^n\phi \to 0$ for $\phi \in D(A^n)$. By a density argument, (6.4) holds for $C$. $\square$

To state the second lemma, we must single out a (non-closed) subspace $\mathcal{M}(A)$. Let $\phi \in \mathcal{H}_{\text{ac}}(A)$. Let $P$ denote the projection onto the cyclic subspace generated by $A$ on $\phi$, that is, the span of $\{e^{-iAs}\phi\}$. Then $\operatorname{Ran} P$ is isomorphic to $L^2(\mathbb{R}, d\nu_\phi)$ in such a way that $\phi$ goes to the function 1 and $e^{-iAs}$ to multiplication by $e^{-ixs}$. Let $P\psi$ correspond to the function $f(x)$ under this correspondence. Then $\int d\nu_\phi = \|\phi\|^2$ and $\int |f(x)|^2\, d\nu_\phi = \|P\psi\|^2 \le \|\psi\|^2$, and

$$(\psi, e^{-iAs}\phi) = (P\psi, e^{-iAs}\phi) = \int \overline{f(x)}\, e^{-ixs}\, d\nu_\phi \tag{6.5}$$

Since $\phi$ is in $\mathcal{H}_{\text{ac}}(A)$, $d\nu_\phi(x) = |G(x)|^2\, dx$ for some $G \in L^2$. Since $fG \in L^2$ by the above, we see by Hölder's inequality that $F = \bar{f}G^2 \in L^1$, a fact used in the proof of Lemma 6.3. Now define

$$\mathcal{M}(A) = \{\phi \mid \phi \in \mathcal{H}_{\text{ac}},\ d\nu_\phi = |G|^2\, dx \text{ and } G \in L^\infty\}$$

We set $|||\phi||| \equiv \|G\|_\infty$. It is easy to see that $\mathcal{M}(A)$ is dense in $\mathcal{H}_{\text{ac}}(A)$. Moreover,

LEMMA 6.4. *If $\phi \in \mathcal{M}(A)$, then for any $\psi$,*

$$\int |\psi, e^{-iAt}\phi)|^2\, dt \le [(2\pi)^{1/2}\|\psi\|\ |||\phi|||]^2 \tag{6.6}$$

PROOF. By (6.5), $(\psi, e^{-iAt}\phi)$ is the Fourier transform of $(2\pi)^{1/2}\bar{f}G^2$ so, by the Plancherel theorem,

$$\int |(\psi, e^{-iAt}\phi)|^2\, dt = 2\pi \int |fG^2|^2\, dx$$

Since $\|G\|_\infty = |||G|||$ and $\int |fG|^2\, dx = \|P\psi\|^2 \le \|\psi\|^2$, the inequality (6.6) follows. $\square$

PROOF OF THEOREM 6.2. Set $W(t) = e^{iAt}Je^{-iBt}$. We want to show that $W(t)\phi$ has a limit for all $\phi$ in $\mathcal{H}_{\text{ac}}(B)$. Since the $\|W(t)\|$ $(= \|J\|)$ are uniformly bounded, we need only prove this for a dense subspace of $\mathcal{H}_{\text{ac}}(B)$, say, for $\phi \in \mathcal{M}(B)$. Since

$$\|W(t) - W(s))\phi\|^2 = (W(t)\phi, (W(t) - W(s))\phi) - (W(s)\phi, W(t) - W(s))\phi)$$

we need only prove that $(\phi, W(t)^*(W(t) - W(s))\phi) \to 0$ as $t, s \to \infty$ or $t, s \to -\infty$. For notational convenience, we consider the case $t \ge s$ and $\to \infty$.

Let $C = AJ - JB$ and introduce the shorthand

$$F_{(a,b)}(Q) = \int_a^b e^{iBw}Qe^{-iBw}\, dw$$

Fix $t, s$ and let

$$G(a) = e^{iaB}W(t)^*W(s)e^{-iaB}$$

Then

$$\begin{aligned}\frac{dG}{da} &= i[e^{iaB}e^{iBt}(BJ^*e^{-iA(t-s)}J - J^*e^{-iA(t-s)}JB)e^{-iBs}e^{-iaB}]\\ &= e^{iaB}Y(t,s)e^{-iaB}\end{aligned}$$

where

$$Y(t,s) = [e^{iBt}C^* e^{-iA(t-s)} J e^{-iBs} - e^{iBt} J^* e^{-iA(t-s)} C e^{-iBs}]$$

Thus

$$W(t)^* W(s) = e^{iaB} W(t)^* W(s) e^{-iaB} - F_{[0,a]}(Y(t,s))$$

so that

$$\begin{aligned}(\phi, W(t)^*(W(t) - W(s))\phi) &= (W(t)e^{-iaB}\phi, (W(t) - W(s))e^{-iaB}\phi) \\ &\quad + (\phi, F_{[0,a]}(Y(t,s) - Y(t,t))\phi)\end{aligned} \tag{6.7}$$

Now, for fixed $t$ and $s$, we do the rather bizarre operation of taking the auxiliary variable $a$ to infinity. Noting that $W(t) - W(s) = i\int_s^t e^{iAw} C e^{-iBw}\, dw$ is compact (since $C \in \mathcal{J}_1 \subset \mathcal{J}_\infty$), we see that the first term goes to zero by Lemma 6.3 whence

$$(\phi, W(t)^*(W(t) - W(s))\phi) = (\phi, F_{[0,\infty)}(Y(t,s) - Y(t,s))\phi) \tag{6.8}$$

where we have used the obvious notation $F_{[0,\infty)} = \lim_{a\to\infty} F_{[0,a]}$, and (6.8) includes the assertion that the limit exists.

Now let $Q(t,s)$ be a family of uniformly bounded operators. We claim that $\lim_{t,s\to\infty}(\phi, F_{[0,\infty)}(Q(t,s)Ce^{-iBs})\phi) = 0$. This claim completes the proof of the theorem since the right side of (6.8) is a sum of terms of this form or their complex conjugates. Let $C = \sum \mu_n(\psi_n, \cdot\,)\eta_n$ be the canonical expansion of $C$. Then

$$\begin{aligned}
&|(\phi, F_{[0,\infty)}(Q(t,s)Ce^{-iBs})\phi)| \\
&= \left|\sum \mu_n \int_0^\infty (e^{-iBa}\phi, Q(t,s)\eta_n)(\psi_n, e^{-iB(a+s)}\phi_n)\, da\right| \\
&\le \left(\sum \mu_n \int_{-\infty}^\infty |(W(t,s)\eta_n, e^{-iBw}\phi)|^2\, dw\right)^{1/2} \left(\sum \mu_n \int_s^\infty |(\psi_n, e^{-iBw}\phi)|^2\, dw\right)^{1/2} \\
&\le \left(\sum \mu_n\right)^{1/2} |||\phi||| \sup_{t,s} \|Q(t,s)\| \left(\sum \mu_n \int_s^\infty |(\psi_n, e^{-iBw}\phi)|^2\, dw\right)^{1/2}
\end{aligned}$$

by Lemma 6.4. By Lemma 6.4 again, $\sum_n \mu_n |(\psi_n, e^{-iBw}\phi)|^2 \in L^1(\mathbb{R})$, so that the integral goes to zero as $s \to \infty$. This completes the proof of the claim, and so of the theorem. □

COROLLARY 6.5 (Kato-Rosenblum Theorem). *If $A - B \in \mathcal{J}_1$, then $\Omega^\pm(A,B)$ exist and are complete.*

PROOF. Taking $J = 1$, $\Omega^\pm(B,A)$ exist, so completeness follows from Proposition 6.1. □

REMARK. This corollary is false if $\mathcal{J}_1$ is replaced by $\mathcal{J}_p$ ($p > 1$); indeed, it is false in a spectacular way. If $\Omega^\pm(A,B)$ exist and are complete, then $A \restriction P_{\mathrm{ac}}(A)$ is unitarily equivalent to $B \restriction P_{\mathrm{ac}}(B)$. On the other hand, given any $A$, even a purely absolutely continuous one, and any $p > 1$, we can find $B$ having only pure point spectrum so that $B - A \in \mathcal{J}_p$. For $p = 2$, this is a result of Weyl [**339**]. Kuroda [**181**] noted that the result extends to any $p > 1$, indeed, to any $\mathcal{J}_\Phi^{(0)}$ with $\Phi$ inequivalent to $\|\cdot\|_1$.

The extraordinary power of adding the $J$ to the statement of Theorem 6.2 is illustrated by

COROLLARY 6.6 (Kuroda-Birman Theorem). *If $A$ and $B$ are self-adjoint and $(A+i)^{-1} - (B+1)^{-1} \in \mathcal{J}_1$, then $\Omega^\pm(A,B)$ exist and are complete.*

PROOF. By symmetry and Proposition 6.1, it suffices to prove that $\Omega^{\pm}(A,B)$ exist. By hypothesis, $AJ - JB \in \mathfrak{I}_1$ if we choose $J = (A+i)^{-1}(B+i)^{-1}$. For $\phi \in D(B)$,

$$\Omega^{\pm}(A,B;(A+i)^{-1}(B+i)^{-1})(B+i)\phi = \Omega^{\pm}(A,B;(A+i)^{-1})\phi \tag{6.9}$$

so, by a density argument, $\Omega^{\pm}(A,B;(A+i)^{-1})$ exist. By hypothesis, $(B+i)^{-1} - (A+i)^{-1} \in \mathfrak{I}_\infty$, so $\Omega^{\pm}(A,B;(B+i)^{-1}-(A+i)^{-1})$ exist (and equal 0) by Lemma 6.3. Thus $\Omega^{\pm}(A,B;(B+i)^{-1})$ exist, so by a formula similar to (6.9), $\Omega^{\pm}(A,B)$ exist. □

Usually the most powerful theorem of this genre is a result of Birman, which requires the notion of:

DEFINITION. Let $A$ and $B$ be self-adjoint operators. We say that $A$ is *subordinate* to $B$ if and only if there exist continuous strictly positive functions $f$ and $g$ on $\mathbb{R}$ so that $\lim_{|t|\to\infty} f(t) = \infty$ and so that $f(A)g(B)^{-1}$ is a bounded operator. If both $A$ is subordinate to $B$ and $B$ is subordinate to $A$, we call them *mutually subordinate.*

REMARK. For example, if $D(A) = D(B)$ or $Q(|A|) = Q(|B|)$, then $A$ and $B$ are mutually subordinate.

Introduce the notation $P_a(A)$ for the spectral projection associated to the interval $[-a,a]$, and $P'_a$ for $I - P_a$:

COROLLARY 6.7 (Birman's Theorem). *Let $A$ and $B$ be self-adjoint operators so that*

(i) *$A$ and $B$ are mutually subordinate.*
(ii) *$P_a(A)(A-B)P_a(B)$ is trace class for any $a < \infty$. Then $\Omega^{\pm}(A,B)$ exist and are complete.*

PROOF. As usual, it suffices to show that $\Omega^{\pm}(A,B)$ exist. Let $J_a = P_a(A)P_a(B)$. Then, by hypothesis, $AJ_a - J_aB \in \mathfrak{I}_1$, so $\Omega^{\pm}(A,B;J_a)$ exist. Let $\phi \in \operatorname{Ran} P_{\rm ac}(B)P_{a_0}(B)$. By a density argument, we need only show that $\Omega^{\pm}(A,B)\phi$ exist. For $a > a_0$, we know that $\Omega^{\pm}(A,B;P_a(A))\phi$ exist, so it suffices to show that

$$\lim_{a\to\infty} [\sup_t \|P'_a(A)e^{-iBt}P_{a_0}(B)\|] = 0 \tag{6.10}$$

Choose $f$ and $g$ in the subordinate definition with $f(A)g(B)^{-1}$ bounded. Let $F(a) = \inf_{|x|>a} f(x)$. Then

$$\begin{aligned}\|P'_a(A)e^{-iBt}P_a(B)\| &\le F(a)^{-1}\|f(A)e^{-iBt}P_{a_0}(B)\| \\ &\le F(a)^{-1}\|f(A)g(B)^{-1}\|\,\|g(B)P_{a_0}(B)\|\end{aligned}$$

so (6.10) holds since $F(a) \to \infty$ as $a \to \infty$. □

In one sense, Birman's theorem is stronger than the others since the hypotheses of either Corollary 6.5 or Corollary 6.6 imply hypothesis (ii) above. (i) is not automatic; for example, if $B = 0$ and $A$ is an unbounded operator with $(A+i)^{-1} \in \mathfrak{I}_1$, we have the hypothesis of Corollary 6.6 obeyed, but $A$ and $B$ are not mutually subordinate (of course, $P_{\rm ac}(A) = P_{\rm ac}(B) = 0$ so the conclusions are trivial in this case).

Birman's theorem is quite powerful. For example, if $B = -\Delta$ and $A-B = \Delta\alpha\Delta$ with $\alpha \ge 0$ and $\alpha$ and its first few derivatives in $L^2_\delta$, $\delta$ large, one can verify the hypotheses of Corollary 6.7 (see, e.g., [**253**]) even though $A - B$ is not a small perturbation of $B$ in any usual sense.

To illustrate the applicability of this theory, let $B = -\Delta$ on $L^2(\mathbb{R}^\nu)$ and $A = -\Delta + V$. There are conditions for the case $\nu = 1, 2$ but one can write down a single condition for $\nu \geq 3$, so we state the theorem in that case:

THEOREM 6.8. *Let $B = -\Delta$ on $L^2(\mathbb{R}^\nu)$ with $\nu \geq 3$. Let $A = -\Delta + V$ where*
(i) *$V \in L^2_\delta$ with $\delta > \nu/2$*
(ii) *$V \in L^{\nu/2} + L^\infty$.*
*Then $\Omega^\pm(A, B)$ exist and are complete.*

PROOF. By hypothesis (i) and Corollary 4.8, $VE_a(B) \in \mathcal{J}_1$ for any $a$. By hypothesis (ii), $V = V_1 + V_2$ with $V_1 \in L^{\nu/2}$, $V_2 \in L^\infty$. By Theorem 4.2, $V_1^{1/2}(B + 1)^{-1/2}$ is in $\mathcal{J}_{p,w}$ $(p = \nu)$ and so compact. It follows that $|\phi, V\phi)| \leq \varepsilon(\phi, B\phi) + c(\phi, \phi)$ so that $Q(A) = Q(B)$. As a result, the hypotheses of Birman's theorem are verified, and so the theorem is proved. $\square$

REMARKS. 1. If $V \in L^p + L^\infty$ with $p > \nu/2$, one can use Theorem 4.1 in place of the subtle Theorem 4.2. If $p > 2$ (and, in particular, if $\nu \geq 5$), $V(B + 1)^{-1}$ is compact, so $D(A) = D(B)$. In any case, one can use Sobolev or Strichartz inequalities (see Chapter 4) in place of compactness results for that part of the proof.

2. One can weaken the hypothesis of this theorem; see [**253**].

Roughly speaking, the hypothesis, $V \in L^2_\delta$ with $\delta > \nu/2$, requires that $V$ has $|x|^{-\nu-\varepsilon}$ fall-off. For many years, it has been known that $\Omega^\pm(-\Delta+V, -\Delta)$ exist if $V$ has $|x|^{-1-\varepsilon}$ fall-off. Completeness in this generality has been proven by Agmon [**3**] and Kuroda [**183, 184**] (see, e.g., [**254**, Section XIII.8]) and also by Enss [**110**] but only using more sophisticated techniques than the trace class method. However, many years ago, Kuroda, in a paper [**182**] which has not received the attention it deserves, showed how the trace class theory could be used to treat $|x|^{-1-\varepsilon}$ fall-off if $V$ is spherically symmetric. His result is:

THEOREM 6.9. *Let $V$ be a function on $\mathbb{R}^\nu$, $\nu \geq 3$, of the form $V(x) = V(|x|)$. Suppose that*

$$\int_0^1 r|V(r)|\,dr + \int_1^\infty |V(r)|\,dr < \infty \tag{6.11}$$

*Then $-\Delta + V$ is a closed quadratic form on $Q(-\Delta)$ and $\Omega^\pm(-\Delta + V, -\Delta)$ exist and are complete.*

SKETCH OF PROOF. (See [**253**, Section XI.4] for further details.) By a spherical harmonic expansion, $L^2(\mathbb{R}^\nu) \cong \oplus L^2(0, \infty; dr)$ in such a way that $-\Delta \cong \oplus h_\ell$ and $-\Delta+V \cong \oplus(h_\ell+V)$ where $h_\ell = d^2/dx^2 + \ell(\ell+1)x^{-2}$ (with $u(0) = 0$ boundary conditions if $\ell = 0$; note $\ell$ may not be an integer if $\nu \neq 3$, but $\ell \geq 0$ so long as $\nu \geq 3$). Suppose we can show that

$$|V|^{1/2}(h_\ell + 1)^{-1}|V|^{1/2} \in \mathcal{J}_1 \tag{6.12}$$

Then $|V|^{1/2}(h_\ell + 1)^{-1/2}$ is compact so $-\Delta + V$ is a closed quadratic form and moreover, $-(h_\ell+V+1)^{-1}+(h_\ell+1)^{-1} = [(h_\ell+V+1)^{-1}(h_\ell+1)^{1/2}][(h_\ell+1)^{-1/2}V(h_\ell+1)^{-1/2}](h_\ell + 1)^{-1/2}$ is trace class, so by the Kuroda-Birman theorem, this theorem is proved. To prove (6.12), we note that $0 \leq (h_\ell + 1)^{-1} \leq (h_0 + 1)^{-1}$, so it suffices to consider the case $\ell = 0$. Now $(h_0 + 1)^{-1}$ is an integral operator with kernel $k(x, y) = e^{-r} \sinh z$ where $r = \max(x, y)$, $z = \min(x, y)$. Thus, *formally*,

$$\operatorname{Tr}(|V|^{1/2}(h_\ell + 1)^{-1}|V|^{1/2}) = \int |V(r)| e^{-r} \sinh r \, dr \tag{6.13}$$

is finite by (6.11). The actual verification of (6.13) uses ideas related to Theorems 2.7(d) and 3.9. □

We close this chapter by showing how the $J$-freedom of Theorem 6.2 and Lemma 6.3 make a powerful combination. The following result was originally proved by Deift-Simon [**88**] using much more sophisticated methods. The proof below follows ideas of Combescure-Ginibre [**78**].

THEOREM 6.10. *Let $V \in L^1(\mathbb{R}^\nu)$ be a non-negative function of compact support. Let $-\Delta + V$ indicate the form sum. Then $\Omega^\pm(-\Delta + V, -\Delta)$ exist and are complete.*

PROOF. Let $\chi$ be a function in $C_0^\infty(\mathbb{R}^\nu)$ which is 1 on $\operatorname{supp}(V)$. Let $J = (1 - \chi)$. Then $(-\Delta + V)J - J(-\Delta) = -(\Delta J) - 2(\nabla J) \cdot \nabla$ so $[(-\Delta + V)J - J(-\Delta)]E_a(-\Delta)$ is trace class for any $a$. Using the fact that $(-\Delta + V)^{-1/2}(-\Delta)^{1/2}$ is bounded, and the method of proof of Birman's theorem, we conclude that $\Omega^\pm(-\Delta + V, -\Delta; J)$ and $\Omega^\pm(-\Delta, -\Delta + V; J)$ exist. Now $(1 - J)(-\Delta + 1)^{-1/2}$ and $(1 - J)(-\Delta + V + 1)^{-1/2} = [(1 - J)(-\Delta + 1)^{-1/2}][-\Delta + 1)^{1/2}(-\Delta + V + 1)^{-1/2}]$ are compact (e.g., by Theorem 4.1), so $\Omega^\pm(-\Delta + V; \Delta; 1 - J) = \Omega^\pm(-\Delta, -\Delta + V; 1 - J) = 0$ exist. □

REMARKS. 1. For more general results of this genre, see [**88**], [**78**], or [**253**, Section XI.4].

2. For further applications of the trace class theory to quantum mechanical scattering, see Combes [**76**], Simon [**299, 302**], and Davies-Simon [**82**].

CHAPTER 7

# Bound State Problems

In this chapter, we want to apply trace ideal methods to study the eigenvalues of $-\Delta + V$. Since $-\Delta + V$ is not compact, we must first rephrase the problem in a way involving compact operators. Recall that $V^{1/2} \equiv V/|V|^{1/2}$.

LEMMA 7.1. *Suppose that $V$ is $-\Delta$-form bounded and there exists $V_n \in C_0^\infty$ so that $\|(-\Delta+1)^{-1/2}(V_n - V)(-\Delta+1)^{-1/2}\| \to 0$ as $n \to \infty$. Then $E < 0$ is an eigenvalue of $-\Delta + V$ if and only if $+1$ is an eigenvalue of*

$$K_E(V) \equiv -V^{1/2}(-\Delta - E)^{-1}|V|^{1/2} \tag{7.1}$$

*in which case their multiplicities are equal.*

PROOF. A direct proof can be found in [**292**]. Here we rely a little more on trace ideal ideas. By a limiting argument, one first shows that one need only consider $V \in C_0^\infty$. In that case, we first write:

$$(-\Delta + V - E)^{-1} = (-\Delta - E)^{-1}(1 + V(-\Delta - E)^{-1})^{-1}$$

to conclude that $E$ is an eigenvalue of $-\Delta + V$ if and only if $-1$ is an eigenvalue of $V(-\Delta - E)^{-1}$. But for $A, B$ bounded, we have $\sigma(AB)\backslash\{0\} = \sigma(BA)\backslash\{0\}$, (see, e.g., [**87**]) so $V(-\Delta - E)^{-1}$ has eigenvalue $-1$ if and only if $-K_E(V)$ has eigenvalue $-1$. The multiplicity question is left to the reader. □

For later purpose we note that the absolute value of $K_E$ is given by

$$|K_E(V)| = |V|^{1/2}(-\Delta - E)^{-1}|V|^{1/2} \tag{7.2}$$

We want to consider two different applications of the reduction to (7.1): First, bounds on

$$N_E(V) = \dim P_{(-\infty,E]}(-\Delta + V)$$

for $E \le 0$, where $P_\Omega(A)$ is a spectral projection. ($N_E(V)$ is the number of bound states of energy less than $E$.) We will denote $N_{E=0}(V)$ by $N(V)$. A general review of these kinds of bounds and other methods of proof may be found in [**250**].) Second, we will use Fredholm theory to obtain perturbation series for the ground state energy (= lowest eigenvalue) in $\nu = 1$ or 2 dimensions.

The general procedure for relating $N(V)$ to eigenvalues of $K_E$ is given by the following result obtained independently by Birman [**34**] and Schwinger [**282**].

PROPOSITION 7.2 (The Birman-Schwinger Principle). *$N_E(V)$ ($E < 0$) is exactly the number of eigenvalues of $K_E$ which are larger than or equal to 1. For $E \le 0$,*

$$N_E(V) \le \#\{\text{eigenvalues of } |K_E(V)| \ge 1\} \tag{7.3}$$

*In particular, for any $p$, we have*

$$N_E(V) \le \|K_E(V)\|_p^p \qquad N_E(V) \le (\|K_E(V)\|_{p,w}^*)^p \tag{7.4}$$

PROOF. Let $\varepsilon_n(\lambda)$ be the $n$-th eigenvalue of $-\Delta+\lambda V$, counting from $\inf \sigma(-\Delta+\lambda V)$ and counting multiplicity. If $-\Delta+\lambda V$ has only $m$ eigenvalues below 0, we set $\varepsilon_n(\lambda)=0$ for $n \geq m+1$. Then the $\varepsilon_n(\lambda)$ are monotone decreasing and continuous for $\lambda \geq 0$. Moreover, once $\varepsilon_n(\lambda)<0$, they are strictly monotone. Thus, for $E<0$,

$$\begin{aligned} N_E(V) &= \#\{\lambda \leq 1 \mid \varepsilon_n(\lambda)=E, \text{ some } n\} \\ &= \#\{\lambda \leq 1 \mid \lambda K_E \text{ has eigenvalue } 1\} \\ &= \# \text{ of eigenvalues of } K_E \geq 1 \end{aligned}$$

From this, we conclude that (7.3) holds for $E<0$. To obtain the $E=0$ result, we note that $|K_E(V)| \leq |K_{E=0}(V)|$ and $N(V)=\lim_{E\uparrow 0} N_E(V)$. □

THEOREM 7.3. (a) *For* $\nu=3$,

$$N(V) \leq (4\pi)^{-2} \int |V(x)|\,|V(y)|\,|x-y|^{-2}\,dx dy \tag{7.5}$$

(b) *Let $V$ be a function on $(0,\infty)$ and let $n_\ell(V)$ denote the number of negative eigenvalues of $-d^2/dx^2+\ell(\ell+1)x^{-2}+V(x)$ (with $u(0)=0$ boundary conditions if $\ell=0$). Then*

$$n_\ell(V) \leq (2\ell+1)^{-1} \int_0^\infty x|V(x)|\,dx \tag{7.6}$$

(*Of course, for spherically symmetric $V$ on $\mathbb{R}^3$, $N(V)=\sum_{\ell=0}^\infty (2\ell+1)n_\ell(V)$.*)

(c) *For* $\nu \geq 3$,

$$N(V) \leq c_\nu \int |V(x)|^{\nu/2}\,d^\nu x \tag{7.7}$$

*for some $\nu$-dependent constant.*

PROOF. (a) By (5.15) with $\alpha=0$, $|K_E(V)|$ has kernel $(4\pi)^{-1}|V(x)|^{1/2}|x-y|^{-1}|V(y)|^{1/2}$ when $E=0$, so that (7.5) is just (7.4) with $p=2$ (Hilbert-Schmidt norm).

(b) Let $h_{0,\ell}=-d^2/dx^2+\ell(\ell+1)x^{-2}$. Then $(h_{0,\ell})^{-1}$ is an integral operator with kernel $(2\ell+1)^{-1}x_<^{\ell+1}x_>^{-\ell}$ where $x_<=\min(x,y)$ and $x_>=\max(x,y)$. As in the proof of Theorem 6.9, $\operatorname{Tr}(|V|^{1/2}h_{0,\ell}^{-1}|V|^{1/2})=(2\ell+1)^{-1}\int_0^\infty x|V(x)|\,dx$, so (7.4) implies (7.6).

(c) Let $Y_E(V)=(-\Delta)^{-1/2}|V(x)|^{1/2}$ so that $K_E(V)=Y_E^*Y_E$. Then, by (7.4),

$$N_E(V) \leq (\|K_E(V)\|_{\nu/2,w}^*)^{\nu/2} = (\|Y_E(V)\|_{\nu,w}^*)^\nu$$

Since $\nu \geq 3$ and $Y_E(V)=g(-i\nabla)|V(x)|^{1/2}$ with $g \in L_w^\nu$, Cwikel's theorem (Theorem 4.2) implies that

$$\|Y_E(V)\|_{\nu,w}^* \leq \left(c_\nu \int |V(x)|^{\nu/2}\right)^{1/\nu}$$ □

REMARKS. 1. (a) is due to Birman [**34**] and Schwinger [**282**] independently. (b) is originally due to Bargmann [**26**] whose proof was very different from the above proof due to Birman and Schwinger. (c) is due independently to Rosenblum [**259**], Lieb [**198**], and Cwikel [**80**]. Lieb, who gets the smallest value for $c_\nu$, uses Wiener path integrals. Rosenblum uses approximation methods in Sobolev spaces pioneered by Birman-Solomjak [**39**] and Birman-Borzov [**35**]. Cwikel uses the method above, which follows the proof of a weaker bound by Simon [**296**].

2. (c) is the most subtle and interesting of these results; for let $\tau_\nu$ be the volume of the unit ball in $\mathbb{R}^\nu$. Then

$$\text{volume in } \mathbb{R}^{2\nu} \text{ of } \{(p,x) \mid p^2 + V(x) < 0\} = \tau_\nu \int (V_-(x))^{\nu/2}\, dx$$

where $V_- = \max(-V, 0)$, so (c) says that, up to a constant, the quasi-classical value $(2\pi)^{-\nu}\text{vol}\{p^2 + V < 0\}$ is a bound on $N(V)$. ($(2\pi)^{-\nu}$ comes from the fact that $\hbar = 1$ so $h = 2\pi$ in the above units.)

3. For sufficiently nice $V$, it can be proven directly [**45, 216, 326**] that $N(\lambda V)/\lambda^{\nu/2} \to \tau_\nu/(2\pi)^\nu \int (V_-(x))^{\nu/2} d^\nu x$ as $\lambda \to \infty$ (see also [**201**]). This and the Birman-Schwinger principle imply that $\mu_n(K_E(V)) \sim n^{-1/\nu}$, so $K_E(V)$ is precisely in $\mathfrak{I}_{\nu,w}$ and not in any smaller space.

4. The asymptotic result of Remark 3 can be extended from nice $V$ to nasty $V$ using bounds like (c). This idea was systematically developed by Birman, Borzov, and Solomjak [**35, 41, 42**] and rediscovered by Simon [**296**].

5. See Lieb-Thirring [**203**] for interesting conjectures on the best value of $c_\nu$ and extensions to inequalities on the moments. Applications to the stability of matter appear in Lieb-Thirring [**202**].

6. From (c), one can deduce bounds on the number of eigenvalues less than $E$ of $-\Delta_\Omega$, the Laplacian in a bounded domain $\Omega \subset \mathbb{R}^\nu$ with Dirichlet boundary conditions; see Lieb [**198**].

This leaves the case $\nu = 1, 2$. First, one has a rather crude result:

PROPOSITION 7.4. *Let $\nu = 1$ or $2$. Let $V \leq 0$ and not identically zero. Then $-\Delta + V$ has a least one negative eigenvalue.*

REMARK. This should be compared with $\nu \geq 3$ where, by Theorem 7.3(c), there are no negative eigenvalues if $\|V\|_{\nu/2}$ is sufficiently small.

PROOF. (Simon [**294**]) Notice that when $V \leq 0$, $K_E(V)$ is a positive compact operator, so $\|K_E(V)\|$ is an eigenvalue. Thus it certainly suffices to prove that

$$\lim_{E\uparrow 0} \|K_E(V)\| = \infty \tag{7.8}$$

((7.8) is also necessary since $\lim_{E\downarrow 0} \|K_E(V)\| = L < \infty$ implies $-\Delta + \lambda V$ has no negative eigenvalues for $\lambda < L^{-1}$.) Thus, we need only find a fixed $\phi$ with $(\phi, K_E(V)\phi) \to \infty$. But

$$(\phi, K_E\phi) = \int \frac{|\widehat{V\phi}(p)|^2}{p^2 - E}\, d^\nu p \tag{7.9}$$

By choosing $\phi$ with $V\phi$ in $L^1$ and $\widehat{V\phi}(0) \neq 0$, we see that the right side of (7.9) diverges as $E \uparrow 0$ for $\nu = 1, 2$. □

REMARK. The fact that $\int_{|p|\leq 1} p^{-2}\, d^\nu p = \infty$ for $\nu = 1, 2$ (infrared divergence) is responsible for a great many phenomena that distinguish $\nu \leq 2$ from $\nu \geq 3$. Besides the above distinction on bound states, there is the recurrence of random walks and Brownian motion if $\nu \leq 2$ [**106**], the absence of continuous symmetry breaking if $\nu \leq 2$ [**220**] and its occurrence if $\nu \geq 3$ [**118**], and the absence of mass zero Gladstone bosons in two space dimensions [**75**].

We want to study these phenomena in more detail, concentrating on the case $\nu = 1$. The fact that (7.9) is divergent only if $\widehat{V\phi}(0) \neq 0$ suggests that a single subtraction is all that is needed to remedy the matter, for example, that only a rank one piece of $K_E$ is producing the trouble. This is indeed the case: Such a decomposition and its consequences was first developed by Simon [**298**]. To extend some of his results, Blankenbecler, Goldberger, and Simon [**47**] provided a clever decomposition, which we use below. Let $L_E(V)$ and $M_E(V)$ be the operators with kernels ($\alpha \equiv \sqrt{-E}$)

$$L_E(V) = -(2\alpha)^{-1} V^{1/2}(x) e^{-\alpha|x|} e^{-\alpha|y|} |V|^{1/2}(y) \tag{7.10}$$

$$M_E(V) = -(2\alpha)^{-1} V^{1/2}(x)[e^{-\alpha|x-y|} - e^{-\alpha(|x|+|y|)}]|V|^{1/2}(y) \tag{7.11}$$

Clearly, $L_E + M_E = K_E$ and $L_E$ is rank one. Moreover, $M_E$ has a pointwise limit as $\alpha \downarrow 0$, so we have at least formally realized the idea that the divergence is due to a rank one piece.

It is very useful to realize $M_E$ in an operator form. Let $h_D$ be the operator $-d^2/dx \oplus -d^2/dx$ on $L^2(0,\infty) \oplus L^2(-\infty,0) = L^2(-\infty,\infty)$ with zero boundary condition at $x = 0$. Then one can easily see that

$$M_E(V) = -V^{1/2}(h_D - E)^{-1}|V|^{1/2} \tag{7.12}$$

(Note, in particular, that $M(x,y) = 0$ if $x < 0 <$ or $y < 0 < x$.) On account of Bargmann's bound (7.6), we have:

THEOREM 7.5 (Blankenbecler-Goldberger-Simon [**47**], Klaus [**169**]). *For* $\nu = 1$,

$$N(V) \leq 1 + \int_{-\infty}^{\infty} |x|\,|V(x)|\,dx \tag{7.13}$$

PROOF. Fix $E < 0$. By the Birman-Schwinger principle, the number of eigenvalues of $M_E$ bigger than 1 is the number of eigenvalues of $h_D + V$ less than $E$. By (7.6), this is dominated by $\int_{-\infty}^{\infty} |x|\,|V(x)|\,dx$. Since $L_E$ is rank one, $K_E$ can have at most $1 + \int_{-\infty}^{\infty} \cdots$ eigenvalues larger than 1. $\square$

REMARKS. 1. Obviously, in (7.13), we can use $\inf_a \int_{-\infty}^{\infty} |x - a|\,|V(x)|\,dx$ in place of $\int_{-\infty}^{\infty} |x|\,|V(x)|\,dx$.

2. It cannot happen that $N(V)$ is dominated by $g(\int_{-\infty}^{\infty} f(V(x))\,dx)$ for any $f, g$; see [**294**].

Now consider the one-parameter family, $-d^2/dx^2 + \lambda V$, for $\lambda$ small. There is at most one negative eigenvalue of $-d^2/dx^2 + \lambda V$ so long as $|\lambda| \int |x|\,|V(x)|\,dx < 1$. It is clearly given by the condition that $K_E$ have $\lambda^{-1}$ as eigenvalue or that $1 - \lambda K_E = 1 - \lambda(M_E + L_E)$ be non-invertible. Now, for all $E < 0$, $\|M_E\| \leq \|M_{E=0}\| \leq \|M_{E=0}\|_1 = \int |x|\,|V(x)|\,dx$. Thus, for $\lambda$ small, $1 - \lambda M_E$ is invertible. Since $1 - \lambda K_E = (1 - \lambda M_E)(1 - \lambda(1 - \lambda M_E)^{-1} L_E)$, we have that $E(\lambda)$ is determined for $\lambda$ small by

$$\det(1 - \lambda(1 - \lambda M_E)^{-1} L_E) = 0 \tag{7.14}$$

Since $L_E$ is rank one and $\det(1 + A) = 1 + \mathrm{Tr}(A)$ for $A$ rank one, (7.14) is equivalent to $\lambda \mathrm{Tr}[L_E(1 - \lambda M_E)^{-1}] = 1$ or

$$G(\alpha,\lambda) \equiv \alpha + \tfrac{1}{2}\lambda(e^{-\alpha|x|}|V|^{1/2},\, (1 - \lambda M_E)^{-1} e^{-\alpha|y|} V^{1/2}) = 0 \tag{7.15}$$

THEOREM 7.6 (Simon [**298**]). (a) *Let* $\int_{-\infty}^{\infty}(1+|x|)|V(x)|\,dx < \infty$. *Then* $-d^2/dx^2 + \lambda V$ *has an eigenvalue* $E(\lambda)$ *for all small* $\lambda > 0$ *if and only if* $\int_{-\infty}^{\infty} V(x)\,dx \le 0$ ($V$ *not identically zero*).

(b) *If, moreover,* $\int |x|^n |V(x)| < \infty$ *for all* $n$*, then* $E(\lambda)$ *has an asymptotic series of all orders at* $\lambda = 0$.

(c) *If, moreover,* $\int e^{a|x|}|V(x)|\,dx < \infty$ *for some* $a > 0$*, then* $E(\lambda)$ *is analytic at* $\lambda = 0$.

(d) *In all cases,*

$$(-E(\lambda))^{1/2} = -\tfrac{1}{2}\lambda \int V(x)\,dx - \tfrac{1}{4}\lambda^2 \int V(x)|x-y|V(y)\,dxdy + o(\lambda^2) \qquad (7.16)$$

SKETCH OF PROOF. The function $G(\alpha,\lambda)$ defined in (7.15) is $C^\infty$ in $\{(\alpha,\lambda) \mid 0 < \alpha;\ \lambda$ real and $|\lambda|$ small$\}$ with the first two derivatives continuous up to the line $\alpha = 0$. The derivatives at $\alpha = 0$, $\lambda = 0$ are summarized by the Taylor series:

$$G(\alpha,\lambda) =$$

$$\left(\alpha + \frac{1}{2}\lambda \int V(x) + \frac{\lambda^2}{4}\int V(x)|x-y|V(y)\,dxdy\right)\left(1 - \lambda \int |x|V(x)\right) + o(\lambda^2 + \alpha^2)$$

By a slight extension of the usual implicit function theorem, $G(\alpha,\lambda) = 0$ will have a (and exactly one) solution $\alpha(\lambda) > 0$ for $\lambda$ small and positive if and only if the formal leading terms are positive, that is, if and only if the right side of (7.16) is positive for $\lambda$ small. If $\int V\,dx < 0$, this is obvious; while if $\int V\,dx > 0$, this cannot occur. If $\int V(x)\,dx = 0$, it automatically happens that

$$\begin{aligned}\int V(x)|x-y|V(y) &= -\lim_{\beta\downarrow 0} \beta^{-1}\int V(x)e^{-\beta|x-y|}V(y)\,dy \\ &= -\lim_{\beta\downarrow 0} 2(V,(h_0+\beta^2)^{-1}V)\end{aligned}$$

is strictly negative, so the right side of (7.16) is positive for $\lambda$ small. This proves (a) and (d). (c) comes from noting that, under the hypothesis, $G(\alpha,\lambda)$ is analytic near $\alpha = \lambda = 0$ so that analyticity of $(-E(\lambda))^{1/2}$ follows from the analytic implicit function theorem. (b) comes from systematically developing $M_E$ in a geometric series using the fact that $E = O(\lambda^2)$ or $0(\lambda^4)$. □

REMARKS. 1. In $\nu = 2$ dimensions, the rank one piece which is responsible for the divergence is $V^{1/2}(x)(-2\pi)^{-1}\log\sqrt{E}\,|V(y)|^{1/2}$ so that if $\int V(x)\,dx < 0$, then $E(\lambda)$ is not analytic at $\lambda = 0$; rather

$$E(\lambda) \sim -\exp\left[4\pi\left(\lambda\int V(x)\,dx\right)^{-1}\right]$$

2. (a) is only proven in [**298**] if $\int(1+|x|)^2|V(x)|\lambda x < \infty$. The stronger result is from [**47**] and [**169**].

3. In [**47**], it is proven that if $V(x) = cx^{-2} + W(x)$ with $|W(x)| \le (1+|x|)^{-2-\varepsilon}$, then (if $\int V(x)\,dx < 0$),

$$\sqrt{-E(\lambda)} \sim \left(-\int V(x)\,dx\right)[\tfrac{1}{2}\lambda + c\lambda^2\log\lambda + O(\lambda^2)]$$

4. If $V(x) \sim -x^{-\beta}$ $(1 < \beta < 2)$ at infinity, $-\Delta + \lambda V$ has infinitely many bound states in $\nu = 1$ or $\nu = 3$ dimensions, but there is a big difference. If $\nu = 3$, all levels

are (for $1 < \beta < 2$) $0(\lambda^{2/(2-\beta)})$, while in one dimension if $\int V\,dx < 0$, all but one have this order but the ground state only has order $\lambda^2$ and is given by [**47, 169**]

$$\sqrt{-E(\lambda)} \sim -\tfrac{1}{2}\,\lambda \int V(x)\,dx + o(\lambda^{\beta-\varepsilon}) \qquad (\varepsilon > 0)$$

5. If $V(x) \sim x^{-1}$ at infinity, $-\Delta + \lambda V$ has infinitely many bound states of $O(\lambda^2)$, but if $\nu = 1$, there is one state of order $\lambda^2(\log\lambda)^2$ at $\lambda = 0$ given by [**21**]

$$\sqrt{-E(\lambda)} = \tfrac{1}{2}\,\lambda\log(\lambda^{-1}) - \tfrac{1}{2}\,\lambda\log(\log(\lambda^{-1})) + O(\lambda)$$

CHAPTER 8

# Lots of Inequalities

In applications of the methods of these lectures, it is often useful to have inequalities involving the trace and determinant at one's disposal. We give here a potpourri of such results. *The proofs will normally be given for the finite-dimensional case* with some remarks on the extension to infinite dimensions. In addition to these types of inequalities, various expressions that $A \mapsto f(A)$ is convex or monotone are useful. These things are beautifully summarized in Ando [**12**]. We also remind the reader that lots of inequalities on singular values can be found in Chapter 1, especially Sections 1.3, 1.6, 1.8, and 1.9.

## 8.1. Golden-Thompson Inequalities

In 1965, Golden [**136**] and Thompson [**327**] independently discovered that

$$\operatorname{Tr}(e^{A+B}) \leq \operatorname{Tr}(e^A e^B) \tag{8.1}$$

One interest of this result is the following: If $A = f(x)$ and $B = g(-i\nabla)$, then (8.1) reads

$$\operatorname{Tr}(e^{f(x)+g(-i\nabla)}) \leq \int e^{f(x)} e^{g(p)}\, dxdp$$

Taking $g(p) = -p^2$ and $f(x) = -V(x)$, we have

$$\operatorname{Tr}(e^{-\Delta+V(x)}) \leq \int e^{-p^2-V(x)}\, dpdx \tag{8.2}$$

This inequality was also obtained by Symanzik [**324**] using other methods. The left side of (8.2) is a quantum partition while the right side is a classical partition function (for subtler inequalities of this genre, see Lieb [**195, 198**] or Simon [**303**]). Since $\operatorname{Tr}(e^A e^B) = \operatorname{Tr}(e^{A/2} e^B e^{A/2})$, (8.1) can be rewritten

$$\|e^{A+B}\|_1 \leq \|e^{A/2} e^B e^{A/2}\|_1$$

In some work on quantum field theory, Segal [**284**] showed that

$$\|e^{A+B}\|_\infty \leq \|e^{A/2} e^B e^{A/2}\|_\infty \tag{8.3}$$

Generalizations of these result were found first by Lenard [**190**] and then Thompson [**328**]. In particular, we follow below an alternative presented by Thompson as an aside in his paper. This approach was rediscovered and emphasized by Deift [**86, 87**], from whom we learned it. There is also a connection with Kato's proof [**166**] of Theorem 4.1.

THEOREM 8.1. *Let $A$ and $B$ be bounded operators with $AB$ self-adjoint. Let $\phi$ be a continuous monotone increasing function on $[0, \infty)$ so that $t \to \phi(e^t)$ is convex. Then*

$$\sum_n \phi(\mu_n(AB)) \leq \sum_n \phi(\mu_n(BA)) \tag{8.4}$$

*In particular,*

$$\|AB\|_p \leq \|BA\|_p \tag{8.4'}$$

PROOF. As in the proof of Theorem 1.15, it suffices to prove that

$$\prod_{n=1}^{N} \mu_n(AB) \leq \prod_{n=1}^{N} \mu_n(BA) \tag{8.5}$$

for such $N$. But (8.5) is equivalent to

$$\|\Lambda^N(AB)\| \leq \|\Lambda^N(BA)\|$$

so it suffices to prove that under the hypotheses of the theorem,

$$\|AB\| \leq \|BA\| \tag{8.6}$$

(8.6) follows from the fact that for $AB$ self-adjoint, $\|AB\| = \sigma_r(AB)$ (where $\sigma_r(\cdot) =$ spectral radius), that $\sigma_r(AB) = \sigma_r(BA)$ (since $\sigma(AB)\backslash\{0\} = \sigma(BA)\backslash\{0\}$), and that $\sigma_r(BA) \leq \|BA\|$. □

COROLLARY 8.2. *Let $C, D \geq 0$, $p \geq 2$. Then*

$$\|CD\|_p \leq \|C^{p/2}D^{p/2}\|_2^{2/p} \tag{8.7}$$

PROOF. Let $A = C^{p/2}$, $B = D^{p/2}$ and let

$$f(\theta) = \|A^\theta B^\theta\|_{2/\theta} \qquad 0 < \theta \leq 1$$
$$f(0) = 1$$

Then $f$ is continuous and (8.7) reads

$$f\left(\theta_0 = \frac{2}{p}\right) \leq f(1)^{\theta_0} f(0)^{1-\theta_0} \tag{8.8}$$

We will prove that

$$f\left(\frac{\theta_1 + \theta_2}{2}\right) \leq f(\theta_1)^{1/2} f(\theta_2)^{1/2} \tag{8.9}$$

from which it follows that $\log f$ is convex, so (8.8) holds. To prove (8.9), let $\theta_1 \leq \theta \leq \theta_2$, $\theta = \frac{1}{2}(\theta_1 + \theta_2)$. Then

$$\begin{aligned} f(\theta)^2 &= \|A^\theta B^{2\theta} A^\theta\|_{1/\theta} \\ &\leq \|A^{\theta_1} B^{\theta_1} B^{\theta_2} A^{\theta_2}\|_{1/\theta} && \text{(by Theorem 8.1)} \\ &\leq \|A^{\theta_1} B^{\theta_1}\|_{2/\theta_1} \|B^{\theta_2} A^{\theta_2}\|_{2/\theta_2} && \text{(by Hölder's inequality)} \\ &= f(\theta_1) f(\theta_2) && \text{(by } \|C^*\|_p = \|C\|_p) \end{aligned}$$

□

REMARKS. 1. This result also follows by applying complex interpolation (Theorem 2.9) to $f(z) = A^z B^z$.

2. If we take $C = f(x)$, $D = g(-i\nabla)$, $f, g \in L^2$ and use (4.8), we obtain a new proof of Theorem 4.1, essentially one of Kato [**166**].

THEOREM 8.3 (Generalized Golden-Thompson Inequality [**328**]). *Let $\phi$ be a continuous monotone increasing, non-negative function on $[0, \infty)$ with $t \to \phi(e^t)$ convex. Let $\phi(C) \equiv \sum_n \phi(\mu_n(C))$. Then for any self-adjoint $A, B$,*

$$\phi(e^{A+B}) \leq \phi(e^{A/2} e^B e^{A/2}) \tag{8.10}$$

*In particular, for* $1 \le p \le \infty$,

$$\|e^{A+B}\|_p \le \|e^{A/2}e^B e^{A/2}\|_p \tag{8.11}$$

*and for* $2 \le p \le \infty$,

$$\|e^{A+B}\|_p \le \|e^A e^B\|_p \tag{8.12}$$

PROOF. Taking $\phi_n(t) = \phi(t^{2n})$, $A(n) = A/2^{n+1}$, $B(n) = B/2^{n+1}$, we have that

$$\begin{aligned}\phi((e^{A(n)}e^{2B(n)}e^{(A(n)})^{2^n}) &= \phi_n(e^{A(n)}e^{2B(n)}e^{A(n)})\\ &\le \phi_n(e^{2A(n)}e^{2B(n)}) \qquad \text{(by Theorem 8.1)}\\ &= \phi_n((e^{A(n-1)}e^{2B(n-1)}e^{A(n-1)})^{1/2})\\ &= \phi((e^{A(n-1)}e^{2B(n-1)}e^{A(n-1)})^{2^{n-1}})\end{aligned}$$

so, by repetition to $n = 0$,

$$\phi((e^{A(n)}e^{2B(n)}e^{A(n)})^{2^n}) \le \phi(e^{A/2}e^B e^{A/2})$$

Taking $n$ to $\infty$ and using the Lie-Trotter formula [**250**],

$$(e^{A/m}e^{2B/m}e^{A/m})^{m/2} \to e^{A+B} \qquad \text{as } m \to \infty$$

we obtain (8.10). From this, (8.11) is immediate and (8.12) is equivalent to $\|e^{2A+2B}\|_{p/2} \le \|e^A e^{2B} e^A\|_{p/2}$, which is (8.11). □

The results from (8.5) can be used to discuss any symmetric convex function, so

THEOREM 8.4. *For any symmetric norming function* $\Phi$ *on* $\mathbb{C}^n$ *and for any* $n \times n$ *symmetric matrices* $A$ *and* $B$, *we have*

$$\Phi(e^{A+B}) \le \Phi(e^{A/2}e^B e^{A/2})$$

The only technical problem with extending the above to the infinite-dimensional case involves the Lie-Trotter formula, since $e^A e^B$ compact requires at least $A$ or $B$ to be unbounded and Trotter's classic result [**332**] then involves some essential self-adjointness assumptions. Kato [**165**] (see [**167**] for extensions to more than two operators) has proven a general Trotter formula for any $-A$ and $-B$ bounded from below. Kato considers $(-A)$ which is defined on a form domain $Q(A)$ not necessarily dense with $A$ self-adjoint on $\bar{Q}$. $e^A$ is the usual operator on $\bar{Q}$ and 0 on $(\bar{Q})^\perp$. If $-A$ and $-B$ are bounded from below, we define $A + B$ on $Q(A) \cap Q(B)$ (which may not be dense even if $Q(A)$ and $Q(B)$ are). Then, if $-A$, $-B$ are bounded below [**165**],

$$\operatorname*{s-lim}_{n\to\infty} (e^{A/n}e^{B/n})^n = e^{A+B}$$

We thus conclude:

THEOREM 8.5. *If* $-A$ *and* $-B$ *are self-adjoint operators bounded from below, then*

$$\|e^{A+B}\| \le \|e^{A/2}e^B e^{A/2}\|$$

*Moreover, if* $\Phi$ *is a symmetric norm inequivalent to* $\|\cdot\|$ *and if* $e^{A/2}e^B e^{A/2} \in \mathfrak{I}_\Phi$, *then* $e^{A+B} \in \mathfrak{I}_\Phi$ *and*

$$\Phi(e^{A+B}) \le \Phi(e^{A/2}e^B e^{A/2})$$

## 8.2. Lieb's Inequalities

We saw in Chapter 1 that $|\cdot|$ on operators *fails* to obey

$$|A+B| \le |A| + |B|$$

On the other hand, the triangle inequality for $\|\cdot\|_1$ implies

$$\mathrm{Tr}(|A+B|) \le \mathrm{Tr}(|A|+|B|)$$

One can ask about other functions, most naturally, whether

$$\det(|A+B|) \le \det(|A|+|B|)$$

This is unfortunately false; for following [**287**], take

$$A = \begin{pmatrix} 1 & \frac{1}{2} \\ \frac{1}{2} & \frac{1}{4} \end{pmatrix} \qquad B = \begin{pmatrix} 0 & 0 \\ 1 & 0 \end{pmatrix}$$

Then $\det(|A+B|) = |\det(A+B)| = \frac{1}{2}$, while $\det(|A|+|B|) = \frac{1}{4}$. Moreover, in this example, $\det(|A^*|+|B^*|) = 1$ so

$$[\det(|A+B|)]^2 \le \det(|A|+|B|)\det(|A^*|+|B^*|) \tag{8.13}$$

(8.13) is one of a large number of inequalities of Lieb [**199**]; we give a simple proof of his result due to Kato [**166**].

DEFINITION. We call a function from $n\times n$ matrices to $C$ *Liebian* if and only if

$$|F(A^*B)|^2 \le F(A^*A)F(B^*B) \tag{8.14a}$$

and

$$0 \le F(A) \le F(B) \quad \text{if} \quad 0 \le A \le B \tag{8.14b}$$

REMARK. Suppose $F$ is Liebian on $\binom{n}{k}\times\binom{n}{k}$ matrices. Then $\tilde F(A) = F(\Lambda^k(A))$ is Liebian on $n\times n$ matrices since $\Lambda^k(A)$ preserves order (most easily shown by noting $0\le A\le B$ if and only if $0\le A$, $0\le B$, and $\|A^{1/2}B^{-1/2}\|\le 1$), adjoints, and products.

EXAMPLES. 1. $F(A) = \mathrm{Tr}(A)$ is Liebian by the fact that $\mathrm{Tr}(\cdot)$ is order preserving and by the Schwarz inequality. More generally, $F(A) = (\phi, A\phi)$ is Liebian. Thus, by the remark, $\mathrm{Tr}(\Lambda^k(A))$ (and $\det(A)$ in particular) and $(\phi_1,\wedge\cdots\wedge\phi_k, \Lambda^k(A)\phi_1\wedge\cdots\wedge\phi_k)$ are Liebian. Note that $\mathrm{Tr}(\Lambda^k(A))$ is the $k$-th elementary symmetric function of the (generalized) eigenvalues.

2. $F(A) = \|A\|$ is clearly Liebian and thus so is $F(A) = \|\Lambda^k(A)\|$ = product of $k$ largest singular values.

3. $\sigma_r(A)$ = spectral radius of $A$ is Liebian: (8.14a) comes from $\sigma_r(A)\le\|A\|$ and Example 2. Thus $\sigma_r(\Lambda^k(A))$, the product of the $k$ largest eigenvalues, is Liebian.

4. If $F$ is Liebian, so is $F_C(A) \equiv f(C^*AC)$ and the supremum of Liebian functions is Liebian. Thus, for example, $\sup\{\mathrm{Tr}(P\Lambda^k(A)P) \mid P \text{ is a rank } \ell \text{ orthogonal projection}\}$ is Liebian. This function is the sum of the $\ell$ largest products of $k$-eigenvalues.

5. See Merris and Dias da Silva [**221**] for other examples.

THEOREM 8.6 (Lieb's Inequality [**199**]). *If $F$ is Liebian, then*

$$\left|F\Big(\sum_1^m a_i^* b_i\Big)\right|^2 \leq F\Big(\sum_1^m a_i^* a_i\Big) F\Big(\sum_1^m b_i^* b_i\Big) \tag{8.15}$$

$$\left|F\Big(\sum_1^m c_i\Big)\right|^2 \leq F\Big(\sum_1^m |c_i|\Big) F\Big(\sum_1^m |c_i^*|\Big) \tag{8.16}$$

*In particular,* (8.13) *holds.*

PROOF. (Kato [**166**]) Let $F$ be Liebian on $n \times n$ matrices. Let $A, B$ take $\mathbb{C}^n$ into $\mathbb{C}^k$. Then it is still true that $|F(A^*B)|^2 \leq F(A^*A)F(B^*B)$. For let $|A| = \sqrt{A^*A}$ which is $n \times n$. Then $A = U|A|$, $B = V|B|$ so $A^*B = |A|U^*V|B|$. Thus, by (8.14a), $|F(A^*B)| \leq F(|B|^2)F(|A|(U^*V)^*U^*V|A|)$. Now, $V^*UU^*V$ is a self-adjoint contraction, so $|A|(U^*V)^*U^*V|A| \leq |A|^2$. Thus, by (8.14b), we have $F(|A|(U^*V)^*U^*V|A|) \leq F(|A|^2)$.

Given $a_1, \dots, a_m$, let $k = nm$ and $A = a_1 \oplus \cdots \oplus a_m$ from $\mathbb{C}^n$ to $\mathbb{C}^k$. Then $A^*B = \sum a_i^* b_i$, so (8.15) holds. Given $c_i$, let $c_i = u_i|c_i|$ be the polar decomposition of $c_i$ and choose $a_i = |c_i|^{1/2}u_i^*$, $b_i = |c_i|^{1/2}$ so $(a_i^*a_i)^2 = u_i|c_i|u_i^*u_i|c_i|u_i^* = u_i|c_i|\,|c_i|u_i^* = c_ic_i^* = |c_i^*|^2$, that is, $a_i^*a_i = |c_i^*|$. Thus (8.15) implies (8.16). □

The following result, proven first by Rotfel'd [**260**] and rediscovered by Seiler-Simon [**287**] using very different methods, served as motivation for Lieb:

COROLLARY 8.7.

$$|\det(1 + A + B)| \leq \det(1 + |A|)\det(1 + |B|)$$

PROOF. Since $|\det(1 + A + B)|^2 \leq \det(1 + |A| + |B|)\det(1 + |A^*| + |B^*|)$ and $\det(1 + |A|) = \det(1 + |A^*|)$ (since $\mu_n(A) \equiv \mu_n(A^*)$), it suffices to consider the case $A, B$ self-adjoint and positive. But then

$$\begin{aligned}\det(1 + A + B) &= \det((1 + A)^{1/2}(1 + (1 + A)^{-1/2}B(1 + A)^{-1/2})(1 + A)^{1/2}) \\ &= \det(1 + A)\det(1 + (1 + A)^{-1/2}B(1 + A)^{-1/2}) \\ &= \det(1 + A)\det(1 + B^{1/2}(1 + A)^{-1/2}B^{1/2}) \\ &\leq \det(1 + A)\det(1 + B)\end{aligned}$$

where we have used $\sigma((1+A)^{-1/2}B(1+A)^{-1/2})\setminus\{0\} = \sigma(B^{1/2}(1+A)^{-1}B^{1/2})\setminus\{0\}$. □

As far as the infinite-dimensional case, one can clearly consider $F$ on $\mathcal{L}(\mathcal{H})$ and even allow $F$ to take the value $+\infty$. For example, if we take $F(A) = \infty$ if $A - I \notin \mathcal{I}_1$ and $F(A) = \det(A)$ if $A - I \in \mathcal{I}_1$, then $F$ is Liebian; all proofs go through.

## 8.3. Peierls-Bogoliubov and Berezin Inequalities

One of the standard tools in quantum statistical mechanics [**262**] is the Peierls-Bogoliubov inequality,

$$\operatorname{Tr}(\phi(A)) \geq \sum \phi((\eta_n, A\eta_n)) \tag{8.17}$$

where $\phi$ is a convex function and $\eta$ is any complete orthonormal set. Apparently unaware of the connection, Berezin [**30**] proved that if $\phi(0) = 0$,

$$\operatorname{Tr}(P\phi(A)P) \geq \operatorname{Tr}(\phi(PAP)) \tag{8.18}$$

for any projection $P$ and convex function $\phi$. Both are non-commutative versions of Jensen's inequality:

LEMMA 8.8. *If $\phi$ is convex and $d\mu$ is a probability measure on $\mathbb{R}$, then*

$$\phi\Big(\int x\,d\mu(x)\Big) \le \int \phi(x)\,d\mu(x) \tag{8.19}$$

PROOF. Let $a = \int x\,d\mu(x)$. By convexity, we can find $\alpha$ with $f(x) - f(a) \ge \alpha(x-a)$ for all $x$. Integrating this inequality with respect to $d\mu(x)$, we obtain (8.19). □

THEOREM 8.9. *Let $A$ be a self-adjoint matrix on $\mathbb{C}^n$ and let $\eta$ be a normalized vector. Then*

$$\phi((\eta, A\eta)) \le (\eta, \phi(A)\eta) \tag{8.20}$$

*for any convex function $\phi$ on an interval $[a,b] \supset \sigma(A)$. Moreover,* (8.17) *and* (8.18) *hold.*

PROOF. (8.20) is just (8.19) with $d\mu$ the spectral measure for $\eta$ (which is $\sum \theta_i \sigma(x - x_i)$ in the finite-dimensional case). (8.17) follows by summing over any orthonormal set. For (8.18), we choose $\eta_1, \dots, \eta_k$ to be an orthonormal basis for $\operatorname{Ran} P$ consisting of eigenvectors for $PAP$. Then $\sum \phi((\eta_i, A\eta_i)) = \operatorname{Tr}(\phi(PAP)) \le \operatorname{Tr}(P\phi(A)P)$. □

COROLLARY 8.10. *Let $A$ be a positive $n \times n$ matrix. Then*

$$\det A \le \prod_{j=1}^{n} a_{jj} \tag{8.21}$$

PROOF. Let $\phi(x) = \log x$. Then, by (8.17),

$$\operatorname{Tr}(\log A) \le \sum \log a_{jj}$$

(8.21) follows by taking exponentials. □

REMARK. Berezin proves the more general result

$$(\det A)^m \le \prod_{k=1}^{\ell} \det(A_k)$$

where $A_k = P_k A P_k$, the determinant is taken on $\operatorname{Ran} P_k$, and the $P_k$ obey $\sum_{k=1}^{\ell} P_k = m\mathbf{1}$.

It is easy to extend all these results, including (8.21), to infinite dimensions; see, for example, [**30**].

## 8.4. Lieb Concavity

The following result was conjectured by Wigner and Yanase [**341, 342**] on the basis of remarks of Dyson. It was first proven by Lieb [**196**] using a complex interpolation argument. The proof we give is essentially due to Uhlmann [**333**], although he embeds his proof in a largely irrelevant framework of quadratic mean interpolation (this framework is of interest in other contexts, however [**233**]):

THEOREM 8.10 (Lieb Concavity). *Fix $X$ an $n \times n$ matrix and $0 \le t \le 1$. The function*

$$f(A, B) = \mathrm{Tr}(X^* A^t X B^{1-t})$$

*on pairs of positive matrices is jointly concave.*

REMARK. Ando [**12**] has noted that Lieb concavity is equivalent to the concavity of the function $A \mapsto A^t \otimes A^{(1-t)}$ from positive operators on $\mathcal{H}$ to positive operators on $\mathcal{H} \otimes \mathcal{H}$.

LEMMA 8.11. *Let $R_1, R_2, S_1, S_2, T_1, T_2$ be positive operators on a Hilbert space so that $[R_1, R_2] = [S_1, S_2] = [T_1, T_2] = 0$ and*

$$R_1 \ge S_1 + T_1 \qquad R_2 \ge S_2 + T_2$$

*Then*

$$R_1^t R_2^{1-t} \ge S_1^t S_2^{1-t} + T_1^t T_2^{1-t} \tag{8.22}$$

*for $0 \le t \le 1$.*

PROOF. Consider first the case $t = \frac{1}{2}$ and let $R = R_1^{1/2} R_2^{1/2}$, etc. Then

$$\begin{aligned}(\phi, (S+T)\psi) &\le (\phi, S_1\phi)^{1/2}(\psi, S_2\psi)^{1/2} + (\phi, T_1\phi)^{1/2}(\psi, T_2\psi)^{1/2} \\ &\le [(\phi, (S_1+T_1)\phi)^{1/2}(\psi, (S_2+T_2)\psi)^{1/2}] \qquad \text{(by Schwarz)} \\ &\le (\phi, R_1\phi)^{1/2}(\psi, R_2\psi)^{1/2}\end{aligned}$$

so that $\|R_1^{-1/2} C R_2^{-1/2}\| \le 1$ where $C = S + T$. By the method of (8.6), $\|R^{-1/2} C R^{-1/2}\| \le 1$; explicitly,

$$\begin{aligned}\sigma(R^{-1/2} C R^{-1/2}) \backslash \{0\} &= \sigma(R_1^{-1/4} C R_2^{-1/2} R_1^{-1/4}) \backslash \{0\} \\ &= \sigma(R_1^{-1/2} C R_2^{-1/2}) \backslash \{0\}\end{aligned}$$

so $\|R^{-1/2} C R^{-1/2}\| = \sigma_r(R^{-1/2} C R^{-1/2}) = \sigma_r(R_1^{-1/2} C R_2^{-1/2}) \le \|R_1^{-1/2} C R_2^{-1/2}\|$.

By the special case $t = \frac{1}{2}$, we can obtain $t = \frac{1}{4}$ and $t = \frac{3}{4}$ by going halfway between 0 and $\frac{1}{2}$ and between 1 and $\frac{1}{2}$. By repetition, we obtain all dyadic rationals, then by continuity, all $t$. □

PROOF OF THEOREM 8.10. Let $\mathcal{H}$ be a family of $n \times n$ matrices with inner product $(X, Y) = \mathrm{Tr}(X^* Y)$. Given positive matrices $A_0, A_1, B_0, B_1$, and $0 \le \theta \le 1$, let $A = \theta A_0 + (1-\theta) A_1$, $B = \theta A_1 + (1 - A_2\theta)$, and

$$\begin{aligned} S_1(X) &= \theta A_0 X & S_2(X) &= \theta X B_0 \\ T_1(X) &= (1-\theta) A_1 X & T_2(X) &= (1-\theta) X B_1 \\ R_1(X) &= AX & R_2(X) &= XB \end{aligned}$$

Then $(X, R_1^t R_2^{1-t} X) \ge (X, S_1^t S_2^{1-t} X) + (X, T_1^t T_2^{1-t} X)$ is equivalent to

$$\mathrm{Tr}(X^* A^t X B^{1-t}) \ge \theta \mathrm{Tr}(X^* A_0^t X B_0^{1-t}) + (1-\theta)\mathrm{Tr}(X^* A_1^t S B_1^{1-t}) \qquad \square$$

The most important application of Lieb concavity is the proof by Lieb-Ruskai [**200**] of the strong subadditivity of entropy; see Lieb [**197**] for further discussion. The following proof is due to Uhlmann [**334**]. Given Hilbert spaces $\mathcal{H}_1$ and $\mathcal{H}_2$, we define the partial trace $\mathrm{Tr}_2$ from $\mathcal{J}_1(\mathcal{H}_1 \otimes \mathcal{H}_2)$ to $\mathcal{J}_1(\mathcal{H}_1)$ by

$$\mathrm{Tr}_{\mathcal{H}_1}(\mathrm{Tr}_2(A)B) = \mathrm{Tr}_{\mathcal{H}_1 \otimes \mathcal{H}_2}(A(B \otimes 1))$$

for any $B \in \mathcal{L}(\mathcal{H}_2)$. Given any $\rho$ on $\mathcal{H}$ with $\rho \geq 0$, $\operatorname{Tr}(\rho) = 1$, we define $S(\rho) = -\operatorname{Tr}(\rho, \log \lambda_n(\rho)$, that is,

$$S(\rho) = \sum_n -\lambda_n(\rho) \log \lambda_n(\rho)$$

(since $0 \leq \lambda_n \leq 1$, $-\lambda_n \log \lambda_n \geq 0$, so the sum converges or diverges to infinity). Given $\mathcal{H} = \mathcal{H}_1 \otimes \mathcal{H}_2$, we think of $\rho_{12}$ on $\mathcal{H}$, and use $\rho_1, \rho_2$ for its partial traces and $S_1(\rho_{12})$ for $S(\rho_1)$, etc.

THEOREM 8.12 (Strong Subadditivity of Entropy [**200**]). *On* $\mathcal{H}_1 \otimes \mathcal{H}_2 \otimes \mathcal{H}_3$,

$$S_{123} + S_2 \leq S_{12} + S_{23}$$

PROOF. (Uhlmann [**334**]) Let $f_t(A, B) = \operatorname{Tr}(A^t B^{1-t})$. By Theorem 8.10, $f_t$ is concave and, clearly, $f_{t=0}$ is linear, so $\frac{df}{dt}\Big|_{t=0}(A, B) = \operatorname{Tr}(B \log A - \log B)$ is concave. On $\mathcal{H}_1 \otimes \mathcal{H}_2$, let $\rho_{12}$ be given. Let $B = \rho_{12}$ and $A = D_2^{-1}(\rho_1 \otimes 1)$ where $d_2 = \dim \mathcal{H}_2$. Then $\rho_{12} \mapsto (A, B)$ is linear, so

$$\rho_{12} \mapsto \operatorname{Tr}(\rho_{12} \log A - \rho_{12} \log \rho_{12}) = S_{12} - S_1 - \log d_2$$

is concave. Replacing $\mathcal{H}_1$ by $\mathcal{H}_1 \otimes \mathcal{H}_2$ and $\mathcal{H}_2$ by $\mathcal{H}_3$, we see that $\rho_{123} \mapsto S_{123} - S_{12}$ is concave. Let $U$ be a unitary on $\mathcal{H}_1$ and let $dU$ be Haar measure on the unitaries on $\mathcal{H}_1$. Then

$$\int U \rho_{123} U^{-1} \, dU = (1 \otimes \rho_{23}) d_1^{-1}$$

By concavity of $S_{123} - S_{12}$ and unitary invariance of $S$,

$$(S_{123} - S_{12})(\rho_{123}) \leq (S_{123} - S_{12})(d_1^{-1}(1 \otimes \rho_{23}))$$

But $S_{123}(d_1(1 \otimes \rho_{23})) = -\log d_1 + S_{23}$, etc., so

$$S_{123} - S_{12} \leq S_{23} - S_2 \qquad \square$$

The results of Section 8.4 extend easily to infinite dimensions (see [**196, 293**]). Lieb concavity and related results are discussed further in Addendum R.

CHAPTER 9

# Regularized Determinants and Renormalization in Quantum Field Theory

In this chapter, we first describe the way of regularizing the determinant so that it can be extended from $\mathcal{I}_1$ to $\mathcal{I}_p$ for suitable $p$ in accordance with the scheme described at the end of Chapter 5; see that chapter for references. Typical is the formula (5.23) for $\det_2$. For general $A \in \mathcal{I}_1$, one can define $\det_2$ according to

$$\det_2(1+A) = \det(1+A)e^{-\mathrm{Tr}(A)} \tag{5.23}$$

As we shall see, the left side of (5.23) extends to all of $\mathcal{I}_2$ even though both $\det(1+A)$ and $\mathrm{Tr}(A)$ will be infinite for suitable $A$ in $\mathcal{I}_2$. This cancellation of infinities suggests a possible connection with renormalization in quantum field theory, a subject we discuss at the end of this chapter.

LEMMA 9.1. *For any $A \in \mathcal{L}(\mathcal{H})$, let*

$$\mathcal{R}_n(A) = \left[(1+A)\exp\left(\sum_{j=1}^{n-1}(-1)^j j^{-1} A^j\right)\right] - 1 \tag{9.1}$$

*Then $\mathcal{R}_n(A) \in \mathcal{I}_1$ for any $A \in \mathcal{I}_n$. Moreover, $A \mapsto \mathcal{R}_n(A)$ is finitely analytic.*

PROOF. Let $g(z) = (1+z)\exp(\sum_{j=1}^{n-1}(-1)^j j^{-1} z^j) - 1$. Since $\sum_{j=1}^{n-1}(-1)^j j^{-1} z^j$ is the beginning of the Taylor series for $\log(1+z)$, we see that $g(z)/z^n$ is an entire function $h(z)$. Thus $g(A) = A^n h(A)$. If $A \in \mathcal{L}(\mathcal{H})$, then $h(A) \in \mathcal{L}(\mathcal{H})$, so if $A \in \mathcal{I}_n$, then $A^n \in \mathcal{I}_1$, and thus $g(A) \in \mathcal{I}_1$. The analyticity is obvious. $\square$

DEFINITION. $\det_n(1+A) = \det(1+\mathcal{R}_n(A))$ for $A \in \mathcal{I}_n$.

THEOREM 9.2. *Let $A \in \mathcal{I}_n$. Then*

(a) $\det_n(1+\mu A) = \prod_{k=1}^{N(A)}[(1+\mu\lambda_k(A))\exp(\sum_{j=1}^{n-1}(-1)^j j^{-1}\lambda_k(A)^j\mu^j)]$

(b) $|\det_n(1+A)| \le \exp(\Gamma_n\|A\|_n^n)$ *for a suitable constant* $\Gamma_n$.

(c) $|\det_n(1+A) - \det_n(1+B)| \le \|A-B\|_n \exp[\Gamma_n(\|A\|_n + \|B\|_n + 1)^n]$

(d) *If $A \in \mathcal{I}_{n-1}$, then*

$$\det_n(1+A) = \det_{n-1}(1+A)\exp\left[(-1)^{n-1}\frac{1}{n-1}\mathrm{Tr}(A^{n-1})\right]$$

*and, in particular, for $A \in \mathcal{I}_1$,*

$$\det_n(1+A) = \det(1+A)\exp\left(\sum_{j=1}^{n-1}(-1)^j j^{-1}\mathrm{Tr}(A^j)\right)$$

(e) $(1+A)$ *is invertible if and only if* $\det_n(1+A) \neq 0$.

REMARKS. 1. (a) is just a canonical product of genus $(n-1)$ and so is quite natural.

2. The only important property lost is $\det(1+A+B+AB) = \det(1+A)\det(1+B)$. Instead, one has, for example,

$$\det{}_2(1+A+B+AB) = \det{}_2(1+A)\det{}_2(1+B)e^{-\mathrm{Tr}(AB)}$$

PROOF. (a) By the spectral mapping theorem for entire functions,

$$\lambda_k(\mathcal{R}_n(A)) = g(\lambda_k(A))$$

whence (a) follows from Theorem 3.7.

(b) For suitable $\Gamma_n$, $|1+g(z)| \le \exp(\Gamma_n|z|^n)$ so that (b) follows from (a) and $\sum|\lambda_j(A)|^n \le \|A\|_n^n$.

(c) follows from (a), according to Theorem 5.1.

(d) is obvious from (a) and Lidskii's equality, (3.2).

(e) is obvious from (a) or, alternatively, from the fact that $1+A$ is invertible if and only if $1+\mathcal{R}_n(A)$ is invertible. □

We will not bother to develop in detail the theory of the function $-A(1+A)^{-1}\det_n(1+A)$ which parallels the discussion in Chapter 5, except to note the formal definition,

$$D_n(A) = -A\,d(\mathcal{R}(A))\exp\biggl(\sum_{j=1}^{n-1}(-1)^j j^{-1}A^j\biggr) \tag{9.2}$$

where $d(A) = (1+A)^{-1}\det(1+A)$.

THEOREM 9.3 (Plemelj-Smithies Formulae for $\det_n$). *For $A \in \mathcal{I}_n$, define $\alpha_n^{(n)}(A), \beta_n^{(n)}$ by $\det_n(1+\mu A) = \sum_{n=0}^\infty \mu^m \alpha_m^{(n)}(A)/m!$ and $D_n(\mu A) = \sum_{n=0}^\infty \mu^{m+1}\beta_m^{(n)}(A)/m!$. Then $\alpha_m^{(n)}$ and $\beta_m^{(n)}$ are given by the formulae of Theorem 5.4 with* $\mathrm{Tr}(A),\dots,\mathrm{Tr}(A^{n-1})$ *replaced by zero.*

PROOF. As in the proof of Theorem 5.4, it suffices to check the formula for $\alpha_m^{(n)}$. This follows as in that proof from Lemma 5.3 if we note that for $\mu$ small,

$$\det{}_n(1+\mu A) = \exp\biggl(\sum_{m=n}^\infty (-1)^{m+1}\frac{1}{m}\mu^m\mathrm{Tr}(A^m)\biggr)$$

on account of Theorem 9.2(a). □

THEOREM 9.4 (Hilbert-Fredholm Formula). *Let $A \in \mathcal{I}_2$ be of the form of* (3.12). *Then* $\det_2(1+A) = \sum_{m=0}^\infty \mu^m\alpha_m^{(2)}(A)/m!$ *where*

$$\alpha_m^{(2)} = \int_a^b dx_1 \dots \int_a^b dx_m\, \widetilde{K}\begin{pmatrix} x_1 & \dots & x_m \\ x_1 & \dots & x_m \end{pmatrix}$$

*and $\widetilde{K}(\cdot)$ is the determinant of the $m\times m$ matrix $\widetilde{K}_{ij} = K(x_i,x_j)(1-\delta_{ij})$.*

REMARKS. 1. This extends to any $\mathcal{I}_2$ integral operator.

2. The process of passing from $K$ to $\widetilde{K}$ involves removing certain permutations from the expansion of det, namely, those with a one cycle in their decomposition into cycles. In general, $\det_n$ removes all permutations with some $j$ cycle $j=1,\dots,n-1$ in their decomposition. This relates the "bubblession" of [**291**] to the renormalized Matthew-Salam-Seiler formalism below.

3. An alternate proof to the one we give is the following: We know that the expansion of the determinant defining $K\left(\begin{smallmatrix} x_1 & \cdots & x_n \\ x_1 & \cdots & x_n \end{smallmatrix}\right)$ followed by the integration will give a sum of multinomials in $\mathrm{Tr}(K^j)$ that must be identical with the sum of monomials that would result if the Plemelj-Smithies formula were expanded. The $\mathrm{Tr}(K)$ terms in the first expansion come precisely from the diagonal terms in $K\left(\begin{smallmatrix} x_1 & \cdots & x_n \\ x_1 & \cdots & x_n \end{smallmatrix}\right)$, so since these are precisely cancelled in modifying the Plemelj-Smithies expansion, they must be precisely cancelled in modifying the Fredholm expansion.

PROOF. By a continuity argument, we need only check the case where $A \in \mathcal{J}_1$. In that case, let $\alpha_m(\lambda)$ be

$$\int_a^b \cdots \int_a^b dx_1 \dots dx_m\, K_\lambda \begin{pmatrix} x_1 & \cdots & x_m \\ x_1 & \cdots & x_m \end{pmatrix}$$

where $(K_\lambda)_{ij} = K(x_i, x_j)(1 - \lambda\delta_{ij})$. Since

$$\frac{\partial}{\partial a_{11}} \det(a_{ij})_{1 \le j,\, j \le n} = \det(a_{ij})_{2 \le i,\, j \le n}$$

one easily sees that

$$\frac{d\alpha_m}{d\lambda} = -m\left(\int_a^b K(x,x)\, dx\right)\alpha_{m-1}(\lambda)$$

so that (as formal power series)

$$\frac{d}{d\lambda}\left(\sum_{m=0}^{\infty} \frac{\alpha_m(\lambda)\mu^m}{m!}\right) = -(\mathrm{Tr}(A)) \sum_{m=0}^{\infty} \frac{\alpha_m(\lambda)}{m!}\, \mu^m$$

and thus

$$\sum_{m=0}^{\infty} \frac{\alpha_m(\lambda)}{m!} = e^{-\lambda \mathrm{Tr}(A)} \det(1 + A)$$

proving the result. □

This completes our exposition of the general theory of regularized determinants. We close this chapter with a description of a formalism for Yukawa field theories. The formal aspects of this development go back to Matthews and Salam [**217**], but it was made into a rigorous tool only with the important paper of Seiler [**285**]. This paper allowed the two-dimensional Yukawa theory introduced and developed by Glimm and Jaffe to be carried to completion by Cooper-Rosen [**79**] and Magnen-Seneor [**211**]. The derivation of the Matthews-Salam-Seiler formulae that we give is due to Seiler-Simon [**289**]. Alternate derivations can be found in Gross [**140**] and McBryan [**218**] who uses the Euclidean Fermi fields of Osterwalder-Schrader [**231**].

The free (relativistic) Bose-field mass $\mu$ in one space dimension can be described as follows (see [**251**, Section X.7] for more details): There is a Hilbert space $\mathcal{H}_B$ and a time zero field $\phi(x)$, $x \in \mathbb{R}$, which is a Hermitian operator-valued distribution, that is, for $f \in \mathcal{S}(\mathbb{R})$, $\int f(x)\phi(x)\, dx$ is an (unbounded) Hermitian operator, a basic Hamiltonian $H_{0,B}$, and a distinguished vector $\Omega_{0,B}$ with

$$(\Omega_{0,B}, \phi(x_1)e^{-s_1 H_{0,B}}\phi(x_2) \dots e^{-s_{n-1}H_{0,B}}\phi(x_n)\Omega_{0,B}) \equiv S_n(x_1, t_1; \dots; x_n, t_n) \quad (9.3)$$

with $t_1 = 0$, $t_2 = s_1$, $t_3 = s_1 + s_2, \dots, t_n = s_1 + \dots + s_{n-1}$ and

$$S_{2n}(y_1, \dots, y_{2n}) = \sum S_2(y_{i_1} - y_{j_1}) \dots S_2(y_{i_n} - y_{j_n})$$

$$S_{n+1} = 0 \tag{9.4}$$

with the sum over all pairings of $\{1, \dots, 2n\}$ into $n$ pairs and

$$S_2(y) = (2\pi)^{-2} \int \frac{e^{ip\cdot y}\, d^2p}{p^2 + \mu^2}$$

(9.4) can be used to define $S_{2n}$ for all $y_1, \dots, y_{2n} \in \mathbb{R}^2$ as a distribution. If we let

$$S_m(f) = \int f(y_1) \dots f(y_m) S_m(y_1, \dots, y_m)\, d^{2m}y$$

then

$$\sum_{m=0}^{\infty} (i\lambda)^m \frac{S_m(f)}{m!} = \exp\Big(-\frac{\lambda^2}{2}\, S_2(f)\Big) \tag{9.5}$$

Since $S_2(f) = (2\pi)^{-1} \int |\widehat{f}(p)|^2 (p^2 + \mu^2)^{-1}\, d^2p \geq 0$, (9.5) defines a positive definite functional on $\mathcal{S}(\mathbb{R}^2)$ so, by Minlos' theorem (see, e.g., [**124, 303**]), there is a measure $d\mu_0$ on $\mathcal{S}'(\mathbb{R}^2)$ so that

$$S_n(y_1, \dots, y_n) = \int T(y_1) \dots T(y_n)\, d\mu_0(T) \tag{9.6}$$

The Gaussian measure $d\mu_0$ is a multidimensional analog of Wiener measure — unlike Wiener measure, it is not supported on nice $T$'s and this can be thought of as the source of the divergence in quantum field theory.

The free (relativistic) Fermi-field of mass $m$ in one space dimension is defined in terms of a Hilbert space $\mathcal{H}_F$, two time zero fields $\psi_\alpha(x)$ ($\alpha = 1, 2$), a basic Hamiltonian $H_{0,F}$, and a distinguished vector $\Omega_{0,F}$ with

$$\begin{aligned}(\Omega_{0,F}, \psi^*_{\alpha_1}(x_1) e^{-s_1 H_{0,F}} \psi_{\beta_1}(y_1) e^{-s_2 H_{0,F}} \dots \psi_{\beta_n}(y_n) \Omega_{0,B})& \\ \equiv S^F_{n;\alpha_1,\beta_1;\dots;\alpha_n;B_n}(x_1, t_1; y_1, \tau_1; \dots; x_n, t_n; y_n, \tau_n)&\end{aligned}$$

with $t_1 = 0$; $\tau_1 = s_1$, $\tau_2 = s_1 + s_2$; $\tau_2 = s_1 + s_2 + s_3, \dots, \tau_n = \sum_{i=1}^{2n-1} s_i$, and

$$S^F_{n;\alpha_i,\beta_i}(a_i, b_i) = \det[S^F_{2;\alpha_i,\beta_i}(a_i - b_j)]$$

with $a_i = (x_i, t_i)$; $b_i = (y_i, \tau_i)$, and $S^F_2(x)$ is a $2 \times 2$ matrix whose exact form will not concern us — we do note that $\widehat{S}^F_2(p)$ behaves as $|p|^{-1}$ as $p \to \infty$ (at least in norm, some expectations go as $|p|^{-2}$ in special directions).

$\widehat{S}^F_2(p)$ is not a Hermitian matrix so there is no hope of obtaining a formula like (9.6) for $\widehat{S}^F_2$ at least with a usual kind of measure (some kinds of non-commutative integration schemes have been advanced). Self-interacting Bose fields were very successfully treated using the methods made available by (9.6) (see, e.g., [**295**]) — the Matthews-Salam formulae involve a method for treating certain Fermi-Boson interactions. We will be somewhat formal manipulating infinite quantities and formal series; points of rigor are discussed in [**289**]. The (spatially cutoff, unrenormalized) Yukawa Hamiltonian is defined on $\mathcal{H}_B \otimes \mathcal{H}_F$ by

$$H = H_{0,B} \otimes 1 + 1 \otimes H_{0,F} + \sum_{\alpha,\beta} \int_a^b \phi(x) \otimes (\psi^*_\alpha(x) A_{\alpha\beta} \psi_\beta(x)) \equiv H_0 + V$$

where $A$ is a special matrix. Let $\Omega_0 = \Omega_{0,B} \otimes \Omega_{0,F}$ and let us try to write an expression for $(\Omega_0, e^{-tH}\Omega_0)$ involving the measure $d\mu_0$ of (9.6). Begin with the formula,

$$(\Omega_0, e^{-tH}\Omega_0) = \sum_{n=0}^{\infty}(-1)^n a_n$$

$$a_0 = (\Omega_0, e^{-tH_0}\Omega_0)$$

$$a_n = \int_{0\le s_i\le t; \sum_1^{n+1} t_i = t} (\Omega_0, e^{-s_1 H_0} V e^{-s_2 H_0} V \dots e^{-s_{n+1} H_0}\Omega_0)\, ds_1 \dots ds_n$$

obtained by iterating the DuHamel formula,

$$e^{-tH} = e^{-tH_0} - \int_0^t e^{-sH_0} V e^{-(t-s)H}\, ds$$

Next, note that if we write $V = \int_a^b V(x)\, dx$, then

$$\begin{aligned} a_n &= \int_{s_1\dots} \int_a^b dx_1 \dots \int_a^b dx_n (\Omega_0, e^{-s_1 H_0} V(x_1) \dots e^{-s_{n+1} H_0}\Omega) \\ &\equiv \int \dots Q(x_1, t_1; \dots; x_n; t_n) \end{aligned}$$

where $t_1 = s_1$, $t_2 = s_1 + s_2, \dots, t_n = s_1 + \dots + s_n$. Since $e^{-tH_0}$, $V(x)$ and $\Omega_0$ are tensor products, $Q$ breaks up into a product of a Bose piece which we can write as $\int T(x_1, t_1) \dots T(x_n, t_n)\, d\mu_0$ (by (9.8)) and a Fermi piece, $\det(A^T.S_F((x_i, t_i) - (x_j, t_j))$ (where $A^T.S_F \equiv \sum_{\alpha,\beta} A_{\alpha\beta} s_{\alpha\beta}$). Noting the symmetry of $Q$, we see that

$$a_n = \int \left\{ \frac{1}{n!} \left[ \prod_{i=1}^n \int_0^t dt_i \int_a^b dx_i\, T(y_1) \dots T(y_n) \det(A^T.S_F(y_i - y_j)) \right] \right\} d\mu$$

with $y = (x, t)$. Looking at this formula and the Fredholm expansion, we conclude that

$$(\Omega_0, e^{-tH}\Omega_0) = \int \det(1 - K(T))\, d\mu_0(T) \tag{9.7}$$

where $K(T)$ is the operator on $L^2((a, b) \times (0, t))$ with kernel

$$T(y)(A^T.S_F)(y - y')$$

$K(T)$ is almost formally of the form of the operators considered in Chapter 4, but with one important difference. $T$ is a.e. w.r.t. $d\mu_0$ a distribution *not* equal to a function! One then proceeds as follows: First, think of $K$ on $L^2(\mathbb{R}^2)$ with kernel $\chi(y)T(y)A^T.S_F(y - y')$, where $\chi$ is the characteristic function of $(a, b) \times (0, t)$. Let

$$\widetilde{K} = (-\Delta + m^2)^{-1/4} K (-\Delta + m^2)^{1/4}$$

Then formally $\det(1 - K) = \det(1 - \widetilde{K})$ since $\sigma(K)\backslash\{0\} = \sigma(\widetilde{K})\backslash\{0\}$. Moreover, since $\widehat{S}_F \sim 1/p$, $\widetilde{K} \sim p^{-1/2} T p^{-1/2}$, which should be better behaved than $K$. We thus use

$$(\Omega_0, e^{-tH}\Omega_0) = \int \det(1 - \widetilde{K}(T))\, d\mu_0(T) \tag{9.8}$$

Since $\widetilde{K}$ is linear in $T$, one can easily compute $\int \operatorname{Tr}(\widetilde{K}^*\widetilde{K})^n\, d\mu_0$ using (9.6) and the explicit formulae for $\widehat{S}_2$ and $\widehat{S}_2^F$. It turns out that [**285**]

$$\int \operatorname{Tr}(\widetilde{K}^*\widetilde{K})\, d\mu_0 = \infty \qquad \int \operatorname{Tr}(\widetilde{K}^*\widetilde{K})^2\, d\mu_0 < \infty$$

So $\widetilde{K}(T)$ is a.e. in $\mathcal{I}_4$ but not a.e. in $\mathcal{I}_2$. In fact, more subtle analysis [**288**] using complex interpolation shows that $\widetilde{K}(T) \in \mathcal{I}_{2+\varepsilon}$ a.e. for all $\varepsilon > 0$. The right side of (9.8) is thus infinite, but that's all right since the left side is also infinite! The Hamiltonian of the theory requires counterterms formally of the form,

$$H_{\text{ren}} = H + \int_0^b (c + d\phi(x) + e(\phi^2(x))\, dx$$

with $c, d, e$ infinite. To make sense of this, one puts in cutoffs and then controls the limit as the cutoffs go away. One way to do this is to prove (9.8) with cutoffs and show that the right side converges as the cutoffs go away. Then

$$(\Omega_0, e^{-tH_{\text{ren}}}\Omega_0) = \int \det(1 - \widetilde{K}(T)) \exp\Big(-\int \chi[c + dT + eT^2]\Big)\, d\mu_0(T)$$

Here $d$ is just chosen so that $\operatorname{Tr}(-\frac{1}{2}\widetilde{K}(T)) = d \int \chi T\, dT$. It is not quite true that $\operatorname{Tr}(-\frac{1}{2}\widetilde{K}(T)^2) = \int \chi(c + eT^2)$, but the difference, $b(T)$, can be shown to be finite a.e. — this is not as hard since $b(T)$ is an explicit quadratic functional in $T$ so, for example, $\int |b(T)|^2\, d\mu_0(T)$ can be computed. Thus

$$(\Omega_0, e^{-tH_{\text{ren}}}\Omega_0) = \int \det{}_3(1 - \widetilde{K}(T))e^{-b(T)}\, d\mu_0(T) \tag{9.9}$$

At this point, it is still non-trivial that the integral on the right side of (9.9) is finite; this was Seiler's major result in [**285**].

CHAPTER 10

# An Introduction to the Theory on a Banach Space

The Hilbert space techniques presented in the first nine chapters are in fairly definitive shape; the same cannot be said of the theory on a general Banach space where, among other problems, there are a multitude of candidates for the analog of $\mathcal{I}_p$. I will give here a brief introduction to some of the notions and problems. We will say nothing about the general Fredholm theory on Banach spaces except to refer the reader to the papers of Grothendieck [**144**], Lezanski [**192**], and Ruston [**269, 270**].

We use $X$ to denote a generic Banach space. The fact that ideal theory in $\mathcal{L}(X)$ is likely to be a zoo is shown already by the famous result of Enflo [**109**] that there exist Banach spaces in which some compact operators, $A$ (i.e., operators $A$ with $A[\{x \mid \|x\| < 1\}]$ precompact), are not the norm limit of finite rank operators. For a review of some aspects of general ideal theory, see [**255**].

Following Grothendieck [**143**], we define the ideal of nuclear operators $N_1(X)$ as follows:

DEFINITION. Let $A$ be a finite rank operator. Define

$$N_1(A) = \inf\biggl\{\sum_{i=1}^m \|\ell_i\|\,\|x_i\| \biggm| \ell_i \in X^*,\, x_i \in X,\, A = \sum_{i=1}^m \ell_i(\cdot)x_i\biggr\}$$

$N_1(X)$ is the completion of the finite rank operators in the norm $N_1(\cdot)$. $N_1(X)$ are called the *nuclear operators*.

$N_1(\cdot)$ is easily seen to be a norm with

$$\|A\|_{\text{op}} \le N_1(A) \tag{10.1}$$

so that $N_1(X)$ can be realized as a vector space of operators. It is also easy to see that $N_1(X)$ is a Banach space and that $N_1(AB) \le \|A\|_{\text{op}} N_1(B)$ (or $N_1(A)\|B\|_{\text{op}}$) so that $N_1(X)$ is a two-sided ideal.

It is not hard to see, using the invariance of trace on finite-dimensional spaces, that $\sum_{i=1}^m \ell_i(x_i)$ is independent of the way that $A$ is decomposed as $\sum_{i=1}^m \ell_i(\cdot)x_i$ so that we can define a linear functional

$$\text{Tr}(A) \equiv \sum_{i=1}^m \ell_i(x_i) \tag{10.2}$$

on the finite rank operators which extends uniquely to $N_1(X)$ since $|\text{Tr}(A)| \le N_1(A)$. It can be shown [**143**] that for any $A \in N_1(X)$ and $\varepsilon > 0$, there is a convergent expansion $A = \sum_{n=1}^\infty \ell_n(\cdot)x_n$ with $\sum_{n=1}^\infty \|\ell_n\|\,\|x_n\| \le N_1(A) + \varepsilon$ and $\text{Tr}(A) = \sum_{n=1}^\infty \ell_n(x_n)$. Not surprisingly,

PROPOSITION 10.1. *For a separable Hilbert space $\mathcal{H}$, $N_1(\mathcal{H}) \equiv \mathcal{I}_1$, including the norm.*

PROOF. We need only identify $N_1(A)$ and $\|A\|_1$ for finite rank operators. The canonical decomposition $A = \sum \mu_n(A)(\phi_n, \cdot\,)\eta_n$ immediately shows that $N_1(A) \leq \|A\|_1$ (for take $x_n = \eta_n$ and $\ell_n = \mu_n(A)(\phi_n, \cdot))$. On the other hand, if $A = \sum_{n=1}^{N}(\gamma_n, \cdot\,)\eta_n$ is an arbitrary finite rank expansion of $A$, then

$$\mathrm{Tr}(|A|) = \sum_{n=1}^{N}(\gamma_n, U^*\eta_n) \leq \sum_{n=1}^{N} \|\gamma_n\|\,\|\eta_n\|$$

so that $\|A\|_1 \leq N_1(A)$. □

A second class of ideals depends on the extension of the notion of singular number to Banach spaces:

DEFINITION. Let $A \in \mathcal{L}(X)$. Then

$$\mu_n(A) = \inf\{\|A - B\| \mid \text{rank } B \leq n-1\}$$

That this agrees with the usual definition when $X = \mathcal{H}$ is a variant of Theorem 1.5.

DEFINITION. (Pietsch [**236**]) $\mathcal{C}_p(X)$ (also called $\ell_p^{\text{app}}$) is the family of operators with $(\sum|\mu_n(A)|^p)^{1/p} = \mathcal{C}_p(A) < \infty$.

THEOREM 10.2 (Pietsch [**236**]). $\mathcal{C}_1(X) \subset N_1(X)$

PROOF. Let $A \in \mathcal{C}_1(X)$. For each $n$, find $A_n$ of rank $2^n - 2$ so that $\|A - A_n\| \leq 2\mu_{(2^n-1)}(A)$. Then

$$\|A_n - A_{n+1}\| \leq 4\mu_{(2^n-1)}(A)$$

By Lemma 10.4 below,

$$\begin{aligned} N_1(A_n - A_{n+1}) &\leq [\mathrm{rank}(A_n - A_{n+1}]4\mu_{(2^n-1)}(A) \\ &\leq 2^{n+4}\mu_{(2^n-1)}(A) \end{aligned}$$

so

$$\sum N_1(A_n - A_{n+1}) \leq 32\sum 2^{n-1}\mu_{(2^n-1)}(A) \leq 32\sum \mu_m(A) < \infty$$

so $A_n$ is Cauchy in $N_1(X)$. Thus $A_n \to B$ for some $B$ in $N_1(X)$. But then $A_n \to B$ in $\mathcal{L}(X)$ (by (10.1)), so $A = B \in N_1(X)$. □

LEMMA 10.3 (Existence of Auerbach Basis). *Let $V$ be a finite-dimensional normed linear space of dimension $n$. Then there exist $x_1, \dots, x_n \in V$ and $\ell_1, \dots, \ell_n \in V^*$ so that $\ell_i(x_j) = \delta_{ij}$ and $\|\ell_i\| = \|x_i\| = 1$.*

PROOF. Fix a basis $z_1, \dots, z_n$ for $V$. For $k_1, \dots, k_n \in (V^*)_1$, the unit ball in $V^*$, let

$$D(k_1, \dots, k_n) = \det(k_i(z_j))$$

Pick $\ell_1, \dots, \ell_n$ maximizing $D$ (by compactness of $V_1^*$) and let $a_{ij} = \ell_i(z_j)$. Since $D$ is clearly not identically zero, $\det(a_{ij}) \equiv \delta \neq 0$, so $a$ has an inverse $b$. Define

$$x_j = \sum_{i=1}^{n} b_{ij} z_i$$

Then $\ell_k(x_j) = \sum_{i=1}^{n} b_{ij}a_{k\ell} = \delta_{kj}$, so since $\|\ell_i\| \leq 1$, we need only show $\|x_i\| \leq 1$ to conclude that $\|\ell_i\| = \|x_i\| = 1$ and thereby conclude the proof.

By the expansion of det$(\cdot)$ in minors and Cramer's rule for $b$ in terms of minors for $a$,

$$D(\ell_1, \dots, \ell_{j-1}, \ell, \ell_{j+1}, \dots, \ell_n) = \sum_{i=1}^{n} \delta b_{ij} \ell(z_i) = \delta \ell(x_j)$$

By the maximizing property of $D(\ell_1, \dots, \ell_n)$, we see that $|\ell(x_j)| \leq 1$ for all $\ell \in V_1^*$, that is, $\|x_j\| \leq 1$. □

LEMMA 10.4. *Let $A$ be a rank $m$ operator on a Banach space $X$. Then*

$$N_1(A) \leq m\|A\|$$

PROOF. Let $V = \mathrm{Ran}(A)$ and let $\ell_i, x_i$ $(i = 1, \dots, n)$ be normalized families in $V$ and $X^*$ (use Hahn-Banach to extend $\ell_i$ from $V$ to $X$) with $\ell_i(x_j) = \delta_{ij}$. Then

$$A\phi = \sum_{i=1}^{m} \ell_i(A\phi)x_i$$

so $A = \sum_{i=1}^{n} (A^*\ell_i)(\cdot)x_i$, so $N_1(A) \leq \sum \|A^*\ell_i\| \, \|x_i\| \leq m\|A\|$. □

Let us mention one last family of ideals, also due to Pietsch [**237**].

DEFINITION. An operator $A \in \mathcal{L}(X)$ is called $p$-summing if and only if

$$\sum \|Tx_n\|^p \leq c\Bigl(\sup_{\|f\|=1,\, f\in X^*} \sum |\langle f, x_n\rangle|^p\Bigr)$$

for any sequence $x_n \in X$. The smallest $c$ in the above is written $[\Pi_p(T)]^p$. The family $\Pi_p(X)$ of $p$-summing operators is an ideal with norm $\Pi_p(\cdot)$.

EXAMPLE. Let us prove that $\Pi_2(\mathcal{H}) = \mathcal{I}_2$ including norms. Let $T \in \mathcal{I}_2$ have the canonical expansion $T = \sum \mu_n(T)(\phi_n, \cdot\,)\psi_n$. Then for any $x_n$,

$$\begin{aligned}\sum \|Tx_m\|^2 &= \sum_{m,n} \mu_n(T)^2 |(\phi_n, x_m)|^2 \\ &\leq \Bigl(\sum_n \mu_n(T)^2\Bigr)\Bigl[\sup_{\|f\|=1} \sum_m |(f, x_m)|^2\Bigr]\end{aligned}$$

so $T \in \Pi_2(\mathcal{H})$ and $\Pi_2(T) \leq \|T\|_2$. Taking $x_n = \phi_m$, we have $\sum |(f, x_m)|^2 \leq \|f\|^2$ by the orthonormality of the $\phi$'s, and obtain $\|T\|_2 \leq \Pi_2(T)$. If $T$ is 2-summing, then $\sum \|T\phi_n\|^2 \leq |\Pi_2(T)|^2$ for any orthonormal set $\phi_n$ when $T$ is Hilbert-Schmidt by Theorem 2.15.

More surprising than the above is:

THEOREM 10.5 (Kwapien [**185**]). $\Pi_p(\mathcal{H}) = \mathcal{I}_2$ *for all* $p$.

On a general Banach space, the $\Pi_p(\cdot)$ are not all equal.

Finally, we want to say something about Weyl's inequality (1.15) and Lidskii's equality (3.2) on general Banach spaces. We begin with an example of Grothendieck:

THEOREM 10.6 (Grothendieck [**143**]). *Let $F(x, y)$ be a continuous function on $[0, 1] \times [0, 1]$. Then the operator*

$$(A_F\phi)(x) = \int F(x, y)\phi(y)$$

*is in* $N_1(L^\infty(0,1))$ *and*

$$N_1(F) \leq \int \sup_x |F(x,y)|\, dy \tag{10.3}$$

PROOF. Let $\chi_{i,n}$ be the characteristic function of the interval $(\frac{i-1}{2^n}, \frac{i}{2^n})$ and let $B$ be the (finite rank) operator with integral kernel $\sum \alpha_{ij}\chi_{i,n}(x)\chi_{j,n}(y)$. We claim that

$$\|B\| \leq N_1(B) \leq \sum_j (\sup_i |\alpha_{ij}|) 2^{-n} \tag{10.4}$$

The first inequality in (10.4) is just (10.1). As for the second inequality, write

$$B = \sum_j \ell_j(\cdot) f_j$$

with $\ell_j(g) = \int \chi_j(y)g(y)\, dy$ and $f_j = \sum_i \alpha_{ij}\chi_{i,n}$. Then $\|\ell_j\|_\infty^* = \|\ell_j\| = 2^{-n}$ and $\|f_j\| = \sup_i |\alpha_{ij}|$.

Now let

$$F^{(n)}(x,y) = \sum_{\substack{1\leq i\leq 2^n\\ 1\leq j\leq 2^n}} F\biggl(\frac{(i-\frac{1}{2})}{2^n}, \frac{(j-\frac{1}{2})}{2^n}\biggr)\chi_{i,n}(x)\chi_{j,n}(y)$$

and $A_n = A_{F^n}$. Then, by (10.4),

$$N_1(A_n) \leq \Bigl(\sum_j \sup_i |F(2^{-n}(i-\tfrac{1}{2}), 2^{-n}(j-\tfrac{1}{2}))| 2^{-n}\Bigr)$$

$$N_1(A_m - A_n) \leq \sup_{z,z'} \bigl\{|F(z) - F(z')| \bigm| |z-z'| \leq 2^{-n-1}\sqrt{2}\,\bigr\}$$

if $m \geq n$ and $A_n \to A$ in the weak $L^1$-topology. Thus, $A$ is in $N_1(L^\infty)$. □

EXAMPLE. (Grothendieck [**143**]) Carleman [**67**] has constructed a function $f$ on $[0,1]$ continuous with $f(0) = f(1)$ so that its Fourier series coefficients $\alpha_n$ obey $\sum |\alpha_n|^p = \infty$ all $p < 2$. Let $F(x,y) = f((x-y) \mid \mod 1)$. Then $A_F$ has eigenvalues $\alpha_n$ so $A_F \in N_1$, but $\sum |\lambda_n(A)|^p = \infty$ for $p < 2$. Weyl's inequality in the form $\sum |\lambda_n(A)| \leq \mathrm{Tr}(|A|)$ fails and, in particular, the sum in Lidskii's equality does not converge absolutely. It is thus usual to say that Lidskii's theorem fails on a general Banach space. In fact, there are examples which show that one cannot hope to recover Lidskii's theorem by a clever summation procedure, namely ([**204**, p. 87]), there is an operator $A \in N_1(\ell_1)$ with $\mathrm{Tr}(A) = 1$ and $A^2 = 0$, so that $\lambda_n(A) = 0$ for all $n$!

In 1972, Markus and Macaev [**214**] proved Lidskii's theorem for $A \in \mathcal{L}(X)$ with $\sum_{n=0}^\infty \mu_n(T)\log[1+\mu_n(T)^{-1}] < \infty$, including the proof of the absolute convergence of the sum of eigenvalues. Johnson et al. [**163**] have proven:

THEOREM 10.7 (Johnson et al. [**163**]). *If* $A \in \mathcal{C}_1(X)$, *then*

$$\sum |\lambda_n(A)| < \infty$$

and

THEOREM 10.8 (Johnson et al. [**163**]). *If* $\sum |\lambda_n(A)| < \infty$ *for any* $A \in N_1(X)$, *then* $X$ *is topologically isomorphic to a Hilbert space; in particular, if* $\mathcal{C}_1(X) = N_1(X)$, *then* $X$ *is isomorphic to a Hilbert space.*

CHAPTER 11

# Borel Transforms, the Krein Spectral Shift, and All That

Our goal here is to review the spectral theory (with applications) of rank one perturbations of positive selfadjoint operators on a separable complex Hilbert space, $\mathcal{H}$:

$$A_\alpha = A + \alpha B \qquad B = (\varphi, \cdot\,)\varphi \tag{11.1}$$

The cynic might feel that I have finally sunk to my proper level. I started with quantum field theory, analysis in infinitely many variables. That was too hard so I switched to the $N$-body Schrödinger equation; but that was too hard so I switched to one-body, then one-dimensional, then discrete one-dimensional. Finally to rank one perturbations — maybe something so easy that I can say something useful! Alas, we'll see even this is hard and exceedingly rich.

I should warn you that (11.1) is somewhat more general than you might think. First, I'll bet you thought that $\varphi$ was a unit vector. It need not be. Big deal, you think — normalize $\varphi$ and renormalize $\alpha$. But $\varphi$ need not be a normalizable vector. We'll consider $B$'s which are rank one but only form bounded perturbations of $A$. In one fell swoop, we've absorbed the theory of variation of boundary conditions for Sturm-Liouville problems on a half-line!

Moreover, as I'll discuss in Section 11.5, $\alpha$ may be infinite.

In some ways, my recent relationship with rank one perturbations reminds me of my relationship with trace ideals fifteen years ago. Then it turned out three distinct problems: Yukawa field theories, scattering theory, and semiclassical bounds on the number of bound states led me to similar mathematics, namely, estimates on the trace ideal properties of certain pseudodifferential operators.

This past year, three distinct research projects:

1. my work, in part with del Rio, Jitomirskaya, and Makarov, on singular spectrum;
2. my trying to understand the work of Aizenman and Molchanov;
3. my work with Gesztesy on trace formulae

all led to rank one perturbations and enriched my own knowledge of the subject. (1)–(3) are the themes of Chapters 12–14 below.

In this chapter, I'll discuss the general theory in a new systematic way; many of the ideas were developed in discussions with Fritz Gesztesy, to whom I am grateful. A key role will be played by

$$F_\alpha(z) = (\varphi, (A_\alpha - z)^{-1}\varphi) \tag{11.2}$$

In terms of a suitable spectral measure $d\mu_\alpha$,

$$F_\alpha(z) = \int \frac{d\mu_\alpha(\lambda)}{\lambda - z} \tag{11.3}$$

so we'll begin with an exhaustive study of the Borel transform of a measure and its relation to the measure in Section 11.1. That section and Section 11.2 are the technical core behind most of what follows.

## 11.1. Borel Transforms of Measures

Throughout, we let $d\mu$ be a (positive) measure on $[a,\infty)$ for some $a > -\infty$ with

$$\int \frac{d\mu(\lambda)}{|\lambda|+1} < \infty \tag{11.4}$$

REMARKS. 1. We'll suppose that $\operatorname{supp}(\mu)$ is bounded from below and the operator $A$ in (11.1) is positive. Much of our theory works without those restrictions but the details are a little simpler with them and they hold in the applications I want to make, so I've made them.

2. For some results when one only knows that $\int \frac{d\mu(\lambda)}{1+\lambda^2} < \infty$, see Aronszajn-Donoghue [**20**].

When (11.4) holds, we can define for $z \in \mathbb{C}\backslash(-\infty,a)$,

$$F(z) = \int \frac{d\mu(\lambda)}{\lambda - z} \tag{11.5}$$

the *Borel transform* of $\mu$ (also called the Stieltjes transform or the Borel-Stieltjes transform). The key issues concern boundary values of $F$ as $z = x + i\epsilon \downarrow x \in \operatorname{supp}(d\mu)$. We begin with the simple stuff away from $\operatorname{supp}(d\mu)$.

THEOREM 11.1. (i) *$F$ is an analytic function on $\mathbb{C}\backslash[a,\infty)$.*
(ii) *$F$ is positive on $(-\infty,a)$ and $\operatorname{Im} F > 0$ if $\operatorname{Im} z > 0$.*
(iii) $\lim_{\gamma\to\infty} F(-\gamma) = 0$.
(iv) $\lim_{\gamma\to\infty} \gamma F(-\gamma) = \int d\mu(\lambda)$ *where $\int d\mu = \infty$ is allowed. Similarly,*

$$\lim_{\gamma\to\infty} \gamma^2 F'(-\gamma) = \int d\mu(\lambda)$$

(v) $\lim_{\gamma\to\infty} \frac{F'(-\gamma)}{F^2(-\gamma)} = (\int d\mu(\lambda))^{-1}$ *where $\int d\mu = \infty$ is allowed with $\infty^{-1} = 0$.*

PROOF. (i) and (ii) are elementary. (ii) uses

$$\operatorname{Im} F(x+i\beta) = \int \frac{\beta}{(x-y)^2+\beta^2}\, d\mu(y) \tag{11.6}$$

(iii) and (iv) follow from the monotone convergence theorem. Actually, by a slightly more involved argument if $|z| \to \infty$ with $|\arg(-z)| < \epsilon$ with $\epsilon > 0$ and small, it is not hard to see that $(-z)F(z) \to \int d\mu$.

If $\int d\mu < \infty$, a similar use of the monotone convergence theorem shows that

$$\gamma^2 F'(-\gamma) = \gamma^2 \int \frac{d\mu(\lambda)}{(\lambda+\gamma)^2} \to \int d\mu(\lambda)$$

and (iv) follows. If $\int d\mu = \infty$, we use the fact that $|1/[zF(z)]| \to 0$ as $|z| \to \infty$ with $\operatorname{Re} z < 0$. Thus, writing

$$\left(\frac{1}{F}\right)'(z) = (2\pi i)^{-1} \int_{|w-z|=|z|/2} (w-z)^{-2} \frac{1}{F(w)}\, dw$$

we see that $(1/F)' \to 0$ but $(1/F)' = -F'/F^2$, implying (v). □

Next, we talk about boundary values of $F$. The function

$$G(x) = \int \frac{d\mu(y)}{(x-y)^2}$$

which is defined for $x \in (-\infty, \infty)$ and takes values in $(0, \infty]$ (note $\infty$ is allowed), will play an important role.

THEOREM 11.2. (i) *$G(x) < \infty$ implies that*

$$\int \frac{d\mu(y)}{|x-y|} < \infty \tag{11.7}$$

(ii) [**92, 138**] *$\{x \mid G(x) = \infty\}$ is a dense $G_\delta$ in* $\operatorname{supp}(d\mu)$.

(iii) *If* (11.7) *holds and, in particular, if $G(x) < \infty$, then $\lim_{\epsilon\downarrow 0} F(x+i\epsilon)$ exists and is real.*

(iv) $\lim_{\epsilon\downarrow 0} \epsilon^{-1} \operatorname{Im} F(x+i\epsilon) \to G(x)$.

(v) *If $G(x) < \infty$, then $\lim_{\epsilon\downarrow 0}(i\epsilon)^{-1}[F(x+i\epsilon) - F(x+i0)] = G(x)$.*

REMARK. A $G_\delta$ is a countable intersection of open sets.

PROOF. (i) is immediate if one notes that

$$\frac{1}{|x-y|} \le \frac{1}{|x-y|^2} + \frac{|x|+2}{|y|+1}$$

since

$$|y| + 1 \le |x| + |x-y| + 1 \le (|x-y|)(|x|+2)$$

if $|x-y| \ge 1$.

(ii) By the lemma below, if $\mu([0,1]) \ge 1$, there is an $x_0 \in [0,1]$ with $\int \frac{d\mu(y)}{|x_0-y|} = \infty$. So, by scaling, if $\mu([a,b]) > 0$, there is an $x_0 \in [a,b]$ with $\int \frac{d\mu(y)}{|x_0-y|} = \infty$. It follows that if $a_0 \in \operatorname{supp}(d\mu)$, then there exists $a_n \to a_0$ with $\int \frac{d\mu(y)}{|a_n-y|} = \infty$. So, by (i), $\{x \mid G(x) = \infty\}$ is dense in $\operatorname{supp}(d\mu)$. It is a $G_\delta$ because if we define $G_n(x) = \int (|x-y|^2 + n^{-1})^{-1}\, d\mu(y)$, then $G_n$ is continuous and

$$\{x \mid G(x) = \infty\} = \bigcap_m \bigcup_n \{x \mid G_n(x) > m\}$$

(iii) By (11.6), for any $\delta$,

$$\operatorname{Im} F(x+i\beta) \le \beta \int_{|x-y|>\delta} |x-y|^{-2}\, d\mu(y) + \int_{|x-y|\le\delta} |x-y|^{-1}\, d\mu(y)$$

so

$$\limsup_{\beta\downarrow 0} \operatorname{Im} F(x+i\beta) \le \limsup_{\delta\downarrow 0} \int_{|x-y|\le\delta} |x-y|^{-1}\, d\mu(y) = 0$$

if (11.7) holds. (11.7) also implies

$$\lim_{\beta\downarrow 0} \operatorname{Re} F(x+i\beta) = \int (y-x)^{-1}\, d\mu(y)$$

by the dominated convergence theorem.

(iv) follows from (11.6) and the monotone convergence theorem.

(v) If $G(x) < \infty$, then, by the proof of (iii),

$$F(x+i0) = \int \frac{d\mu(y)}{y-x}$$

so

$$(i\epsilon)^{-1}[F(x+i\epsilon)-F(x-i0)] = \int \frac{d\mu(y)}{(y-x)(y-x-i\epsilon)} \to G(x)$$

by the dominated convergence theorem if $G(x)<\infty$. $\square$

LEMMA 11.3. *Let $\mu([0,1])\geq 1$. Then, there exists $x_0\in[0,1]$ with*

$$\int \frac{d\mu(y)}{|x_0-y|} = \infty$$

PROOF. Define $I_n$ inductively as follows. Let $I_1$ be one of $[0,\frac{1}{2}],[\frac{1}{2},1]$ with $\mu(I_1)\geq\frac{1}{2}$. Now subdivide $I_1$ in half and choose $I_2$ to be one of those halves with $\mu(I_2)\geq\frac{1}{4}$. By induction,

$$I_1\supset I_2\supset\cdots \quad \text{all closed} \qquad \mu(I_n)\geq 2^{-n} \qquad |I_n|=2^{-n}$$

By compactness, $\cap I_n=\{x_0\}$ for some $x_0$. Obviously,

$$\int_{I_n} |x-x_0|^{-1}\,d\mu(x) \geq (2^{-n})^{-1}2^{-n}=1$$

But if $\int|x-x_0|^{-1}\,d\mu(x)<\infty$, $\int_{I_n}|x-x_0|^{-1}\,d\mu(x)\downarrow 0$ by monotone convergence, so we must have $\int|x-x_0|^{-1}\,d\mu(x)=\infty$. $\square$

THEOREM 11.4. $\lim_{\beta\downarrow 0}F(x+i\beta)$ *exists and is finite for a.e. $x$.*

PROOF. This is a standard harmonic analysis result on nontangential boundary values of analytic maps (see [**168**]), so we'll only sketch the details. Let $g(z)=(z-i)/(z+i)$, the fractional linear map of the upper half-plane to the unit disk. Thus $\tilde{F}=g\circ F\circ g^{-1}$ is a map of the unit disk to itself. Such maps have boundary values by a standard maximal function argument; see Katznelson [**168**]. By the same argument as in the next theorem, $\{\theta\mid \lim_{r\uparrow 1}\tilde{F}(re^{i\theta})=-1\}$ has measure zero, so $\lim_{\beta\downarrow 0}F(x+i\beta)=\infty$ on a set of measure 0, and so $F$ is finite a.e. $\square$

THEOREM 11.5. *Let $F_1,F_2$ be the Borel transforms of two (positive) measures (one may be zero) and fix $\alpha\in\mathbb{C}$. Then*

$$|\{x\mid F_1(x+i0)-F_2(x+i0)=\alpha\}|>0$$

*only if $\alpha=0$ and $F_1=F_2$.*

PROOF. Let $g$ be defined as in the proof of Theorem 11.4 and let

$$h=e^{iF_1\circ g^{-1}}-(e^{iF_2\circ g^{-1}})e^{i\alpha}$$

$h$ is a bounded analytic function on the disk and the theorem asserts that under $h\not\equiv 0$, $\{\theta\mid\lim_{r\uparrow 1}h(re^{i\theta})=0\}$ has Lebesgue measure zero. This is a standard theorem; see Katznelson [**168**]. $\square$

Let $d\mu_{\text{ac}},d\mu_{\text{sc}},d\mu_{\text{pp}}$ be the absolutely continuous, singular continuous, and pure point parts of a measure $d\mu$. Let $d\mu_{\text{sing}}=d\mu_{\text{sc}}+d\mu_{\text{pp}}$.

The main theorem on Borel transforms is

THEOREM 11.6. *Let $F(z)$ be the Borel transform of a measure $d\mu$ obeying* (11.4). *Then*

(i) $\pi^{-1} \operatorname{Im} F(E+i\epsilon)\, dE \to d\mu$ *weakly in the sense that*

$$\int \pi^{-1} f(E) \operatorname{Im} F(E+i\epsilon)\, dE \to \int f(E)\, d\mu(E) \tag{11.8}$$

*for all continuous functions, $f$, of compact support.*

(ii) $\mu_{\text{sing}}$ *is supported on* $\{E \mid \lim_{\epsilon\downarrow 0} \operatorname{Im} F(E+i0) = \infty\}$.

(iii) $\mu(\{E_0\}) = \lim_{\epsilon\downarrow 0} \epsilon \operatorname{Im} F(E_0+i\epsilon)$. *Moreover, for any* $E_0 \in \mathbb{R}$,

$$\lim_{\epsilon\downarrow 0} \epsilon \operatorname{Re} F(E_0+i0) = 0$$

(iv) $d\mu_{\text{ac}}(E) = \pi^{-1} \operatorname{Im} F(E+i0)\, dE$.

PROOF. (i) It is a well-known calculation that for $f$ continuous of compact support,

$$\lim_{\epsilon\downarrow 0} \frac{1}{\pi} \int f(E') \frac{\epsilon}{(E-E')^2+\epsilon^2}\, dE' \to f(E)$$

This and the dominated convergence theorem easily imply the result.

(ii) We will prove the weaker fact that $\mu_{\text{sing}}$ is supported on $\{E \mid \limsup \operatorname{Im} F(E+i2^{-n}) = \infty\}$. The stronger fact (a theorem of de la Vallée-Poussin) can be found in Saks [**276**] (see Addendum J), but for most applications, this weaker fact suffices. Since the stronger result is true and more elegant, we'll use it. For the theorem of Aronszajn on mutual singularity (Theorem 11.2(iv)), one does need the stronger fact.

We will prove that for any $a$,

$$\bigl\{E \mid \sup_n \operatorname{Im} F(E+i2^{-n}) \le a\bigr\} = A_a$$

has $\mu_{\text{sing}}$-measure 0 from which the result follows. Notice that

$$\begin{aligned}
\operatorname{Im} F(x+i\epsilon) &= \int \frac{\varepsilon}{(x-y)^2+\epsilon^2}\, d\mu(y) \\
&\ge \frac{1}{2\epsilon} \int_{|x-y|\le\epsilon} d\mu(y)
\end{aligned}$$

So for fixed $x \in A_a$,

$$\sup_n 2^n \mu([x-2^{-n}, x+2^{-n}]) \le 2a$$

and so, since any $2^n \le \epsilon^{-1} \le 2^{n+1}$ for some $n$,

$$\sup_n \epsilon^{-1} \mu([x-\epsilon, x+\epsilon]) \le 4a$$

and thus looking at the two subintervals,

$$\sup_{x\in[\alpha,\beta]} |\beta-\alpha|^{-1} \mu([\alpha,\beta]) \le 8a$$

By regularity of measures, for any $\epsilon$, we can find a closed set $C$ and open $O$ with $C \subset O$, so $\mu_{\text{sing}}(\mathbb{R}\backslash C) \le \epsilon$ and $|O| \le \epsilon$. We write $O = \cup_{n=1}^\infty (\alpha_n, \beta_n)$ as a disjoint union of intervals (union of connected components). Then

$$\begin{aligned}
\mu_{\text{sing}}(C \cap A_a) &\le \mu_{\text{sing}}\Biggl(\bigcup_{\{n \mid (\alpha_n,\beta_n)\cap A_a \neq \emptyset\}} (\alpha_n, \beta_n)\Biggr) \\
&\le \sum_{\{n \mid (\alpha_n,\beta_n)\cap A_a \neq \emptyset\}} \mu(\alpha_n, \beta_n) \le 8a \sum_n |\beta_n - \alpha_n| = 8a|O| \le 8a\epsilon
\end{aligned}$$

so

$$\mu_{\rm sing}(A_a) \le \mu_{\rm sing}(C \cap A_a) + \mu_{\rm sing}(\mathbb{R}\backslash C) \le \epsilon + 8a\epsilon$$

Since $\epsilon$ is arbitrary, $\mu_{\rm sing}(A_a) = 0$, as was to be proven.

(iii) The first part is a trivial consequence of the monotone convergence theorem, if we note that $\epsilon^2/(x^2+\epsilon^2)$ converges monotonically to 0 (resp. 1) if $x \neq 0$ (resp. $x = 0$). For the second,

$$\epsilon \int_{|x-y|\ge 1} \frac{|x-y|}{|x-y|^2+\epsilon^2}\, d\mu(y) \le \epsilon \int_{|x-y|\ge 1} \frac{1}{|x-y|}\, d\mu(y) \to 0$$

while

$$\lim_{\epsilon\downarrow 0} \int_{|x-y|<1} \frac{\epsilon|x-y|}{(x-y)^2+\epsilon^2}\, d\mu(y) = 0$$

by the dominated convergence theorem since $\alpha\beta/(\alpha^2+\beta^2) \le \frac{1}{2}$.

(iv) Write $d\mu_{\rm ac}(E) = h(E)\, dE$ and let $F_{\rm ac}(z)$ be the Borel transform of $d\mu_{\rm ac}$. By the Lebesgue theorem on the derivatives of the integral (that says if $h \in L^1$, for a.e. $x$, $h(x) = \lim_{a\to 0}(2a)^{-1}\int_{x-a}^{x+a} h(y)\, dy$; see Katznelson [**168**]), one sees that $\frac{1}{\pi}\operatorname{Im} F_{\rm ac}(E+i0) = h(E)$ for a.e. $E$. By Fatou's lemma, for $f \ge 0$ and continuous,

$$\begin{aligned}\pi^{-1}\int [\operatorname{Im} F(E+i0)] f(E)\, dE &\le \liminf \pi^{-1}\int \operatorname{Im} F(E+i\epsilon) f(E)\, dE \\ &= \int f(E)\, d\mu(E)\end{aligned}$$

Thus $\pi^{-1}\operatorname{Im} F(E+i0)\, dE \le d\mu$, and so for a.e. $E$,

$$\frac{1}{\pi}\operatorname{Im} F(E+i0) \le h(E) = \frac{1}{\pi}\operatorname{Im} F_{\rm ac}(E+i0) \le \frac{1}{\pi}\operatorname{Im} F(E+i0)$$

since $d\mu \ge d\mu_{\rm ac}$ implies $\operatorname{Im} F \ge \operatorname{Im} F_{\rm ac}$. Thus

$$h(E) = \frac{1}{\pi}\operatorname{Im} F(E+i0)$$

as was to be proven. □

## 11.2. Rank One Perturbations: The Set-Up and Basic Formulae

Let $A \ge 0$ be a possibly unbounded selfadjoint operator on a Hilbert space $\mathcal{H}$. One defines the scale of spaces, $\mathcal{H}_s(A)$, associated to $A$ as follows. For $s \ge 0$, $\mathcal{H}_s(A)$ is $D(A^{s/2})$ with the norm

$$\|\varphi\|_s = \|(A+1)^{s/2}\varphi\| \tag{11.9}$$

in which $\mathcal{H}_s$ is easily seen to be complete. For $s < 0$, take $\mathcal{H}$ with the norm given by (11.9) and complete it. In a natural way, $\mathcal{H}_s$ and $\mathcal{H}_{-s}$ are duals in such a way that $\varphi \in \mathcal{H}$ is associated to the functional on $\mathcal{H}_s$ given by $\ell_\varphi(\eta) = (\varphi, \eta)$, so $|\ell_\varphi(\eta)| \le \|\varphi\|_{-s}\|\eta\|_s$ and $\mathcal{H}_s \subset \mathcal{H}_t$ if $s > t$.

$(A+1)^{-t/2}$ is an isometry from $\mathcal{H}_s$ to $\mathcal{H}_{s+t}$, so for $s > 0$, if $\Delta$ is a bounded subset of $\mathbb{R}$, $E_\Delta(A)$, the spectral projection of $A$, maps $\mathcal{H}_{-s}$ to $\mathcal{H}_s$ since $(A+1)^{+s/2}E_0(A)(A+1)^{s/2}$ is bounded. In particular, we can define a spectral measure $d\mu_A^\varphi$ for any $\varphi \in \mathcal{H}_{-s}(A)$ by

$$\mu_A^\varphi(\Delta) = (\varphi, E_\Delta(A)\varphi)$$

It is easy to see that

$$\int \frac{d\mu_A^\varphi(x)}{(|x|+1)^s} < \infty \tag{11.10}$$

and, in particular, the spectral measures of $\varphi \in \mathcal{H}_{-1}(A)$ have Borel transforms.

Now let $\varphi \in \mathcal{H}_{-1}(A)$ and let $b$ be the quadratic form on $\mathcal{H}_{+1}(A)$ given by

$$b_\varphi(\eta, \eta') = \overline{\ell_\varphi(\eta)}\, \ell_\varphi(\eta')$$

$b_\varphi$ is positive but if $\varphi \notin \mathcal{H}$, it is neither closed nor closable.

See [**164, 251**] for a discussion of forms and their perturbation theory.

PROPOSITION 11.7. *For any $\varphi$, $b_\varphi$ is a form bounded perturbation of $A$ with relative bound zero.*

PROOF. Given $\epsilon$, find $\varphi_\epsilon \in \mathcal{H}_{-1}(A)$ so $\|\varphi_\epsilon\|_{-1}^2 \le \frac{1}{2}\epsilon$ and $\psi_\epsilon \equiv \varphi - \varphi_\epsilon \in \mathcal{H}$. Then

$$b_\varphi = b_{\varphi_\epsilon+\psi_\epsilon} \le b_{\varphi_\epsilon+\psi_\epsilon} + b_{\varphi_\epsilon-\psi_\epsilon} = 2b_{\varphi_\epsilon} + 2b_{\psi_\epsilon}$$

So, for any $\eta \in \mathcal{H}_{+1}(A)$,

$$\begin{aligned} b_\varphi(\eta,\eta) &\le 2\|\varphi_\epsilon\|_{-1}^2 \|\eta\|_{+1}^2 + 2\|\psi_\epsilon\|^2 \|\eta\|^2 \\ &\le \epsilon(\eta, A\eta) + (2\|\psi_\epsilon\|^2 + \epsilon)\|\eta\|^2 \end{aligned}$$

as was to be proven. □

The perturbation theory of forms [**164, 251**] thus implies that for any $\alpha \in \mathbb{R}$, $A + \alpha b$ is the quadratic form of a selfadjoint operator $A_\alpha$ with $\mathcal{H}_s(A_\alpha) = \mathcal{H}_s(A)$ if $|s| \le 1$. We will write

$$A_\alpha = A + \alpha B = A + \alpha(\varphi, \cdot\,)\varphi \tag{11.11}$$

although there is no operator $B$ or vector $\varphi$ in the usual sense if $\varphi \notin \mathcal{H}$. We will occasionally write $\|\varphi\|^2$ which may be infinite if $\varphi \notin \mathcal{H}$.

Since $\mathcal{H}_{-1}(A_\alpha) = \mathcal{H}_{-1}(A)$ and so $(A_\alpha - z)^{-1}\varphi \in \mathcal{H}_{+1}(A_\alpha) = \mathcal{H}_{+1}(A) \subset \mathcal{H}$. The strong resolvent formula holds, namely,

$$(A_\alpha - z)^{-1} - (A - z)^{-1} = -\alpha((A_\alpha - \bar{z})^{-1}\varphi, \cdot\,)(A - z)^{-1}\varphi \tag{11.12}$$

Don't blink now because in just about one page we're going to present four of the five critical formulae in the subject (the fifth is (11.17) below).

Define

$$F_\alpha(z) = (\varphi, (A_\alpha - z)^{-1}\varphi) \equiv \int \frac{d\mu_\alpha(x)}{x - z}$$

where $d\mu_\alpha \equiv d\mu_{A_\alpha}^\varphi$ is the spectral measure for $\varphi$ which has a Borel transform by (11.10) and $\mathcal{H}_{-1}(A_\alpha) = \mathcal{H}_{-1}(A)$. We define $F(z) \equiv F_{\alpha=0}(z)$.

Taking matrix elements of (11.12) with $\varphi$ on both sides, we get $F_\alpha(z) - F(z) = -\alpha F_\alpha(z) F(z)$ or, solving for $F_\alpha$, we get the Aronszajn-Krein formula:

$$F_\alpha(z) = \frac{F(z)}{1 + \alpha F(z)} \tag{11.13}$$

the first critical formula. Applying (11.12) to $\varphi$ and using (11.13) to see that $1 - \alpha F_\alpha = [1 + \alpha F(z)]^{-1}$, we get the second key formula:

$$(A_\alpha - z)^{-1}\varphi = (1 + \alpha F(z))^{-1}(A - z)^{-1}\varphi \tag{11.14}$$

Plugging this into (11.12), we get the third key formula:

$$(A_\alpha - z)^{-1} = (A - z)^{-1} - \frac{\alpha}{1 + \alpha F(z)} ((A - \bar{z})^{-1}\varphi, \cdot)(A - z)^{-1}\varphi \tag{11.15}$$

Notice that (11.14) says that for $\eta, z$ fixed, $(\eta, (A_\alpha - z)^{-1}\varphi) = a/(\alpha + b)$, the key to the Aizenman-Molchanov theory.

For the final formula, (11.12) says that $(A_\alpha - z)^{-1} - (A - z)^{-1}$ is rank one, so trace class and, by (11.15), we have

$$\mathrm{Tr}[(A - z)^{-1} - (A_\alpha - z)^{-1}] = \frac{\alpha}{1 + \alpha F(z)} (\varphi, (A - z)^{-2}\varphi)$$

Notice that $(\varphi, (A - z)^{-2}\varphi) = \frac{d}{dz} F(z)$. Thus, we have what I'll call the trace formula, the key to our discussion of the Krein spectral shift in Section 11.4 below:

$$\mathrm{Tr}[(A - z)^{-1} - (A_\alpha - z)^{-1}] = \frac{d}{dz} \log(1 + \alpha F(z)) \tag{11.16}$$

where we take the branch of the log which is positive for $z$ real and very negative (recall, by Theorem 11.1(iii), that even for $\alpha < 0$, $1 + \alpha F(z) > 0$ for $z$ real and very negative).

## 11.3. The Integral Formula

The critical last formula of the general theory is

THEOREM 11.8.

$$\int_{-\infty}^{\infty} [d\mu_\alpha(E)]\, d\alpha = dE \tag{11.17}$$

*in the sense that if $f$ is in $L^1(\mathbb{R}, dE)$, then $f \in L^1(\mathbb{R}, d\mu_\alpha)$ for a.e. $\alpha$, $\int f(E)\, d\mu_\alpha(E) \in L^1(\mathbb{R}, d\alpha)$, and*

$$\int \left( \int f(E)\, d\mu_\alpha(E) \right) d\alpha = \int f(E)\, dE \tag{11.18}$$

PROOF. By an elementary argument, it suffices to prove this result for $f_z(E) = (E - z)^{-1} - (E + i)^{-1}$ for any $z \in \mathbb{C}\backslash\mathbb{R}$ (use analyticity in $z$ and then Stone-Weierstrass approximation). By closing the contour in the upper half-plane

$$\int_{-\infty}^{\infty} f_z(E)\, dE = \begin{cases} 0 & \text{if } \mathrm{Im}\, z < 0 \\ 2\pi i & \text{if } \mathrm{Im}\, z > 0 \end{cases}$$

On the other hand, by (11.13),

$$\begin{aligned} \int f_z(E)\, d\mu_\alpha(E) = F_\alpha(z) - F_\alpha(-i) &= \frac{1}{\alpha + F(z)^{-1}} - \frac{1}{\alpha + F(-i)^{-1}} \\ &\equiv h_z(\alpha) \end{aligned}$$

Now, if $\pm \mathrm{Im}\, z > 0$, then $\pm \mathrm{Im}\, F(z) > 0$, so $\pm \mathrm{Im}\, F(z)^{-1} < 0$. Thus, $h_z(\alpha)$ has either two poles in the lower half-plane if $\mathrm{Im}\, z < 0$ or one in each half-plane if $\mathrm{Im}\, z > 0$. The same contour integral as for $E$ (but now for $\alpha$!) implies that

$$\int_{-\infty}^{\infty} d\alpha\, h_z(\alpha) \begin{cases} = 0 & \text{if } \mathrm{Im}\, z < 0 \\ \equiv 2\pi i & \text{if } \mathrm{Im}\, z > 0 \end{cases}$$

so (11.18) is proven for $f_z(E)$. □

The above implies a regularity result for the density of states in $n$-dimensional Jacobi matrices (see Theorem 11.19 below) due to Wegner [**337**]. Wegner proved his result by tracking eigenvalues in finite-dimensional approximations, but his underlying result is similar to (11.17) so this is sometimes called a Wegner estimate. The Russian literature, however, has similar results much earlier in the case of boundary condition dependence; see Javrjan [**161**].

## 11.4. The Krein Spectral Shift

In this section, we make a presentation of the Krein spectral shift [**178**] due to Gesztesy and Simon. By the general theory of Borel transforms, $F(z)$ has a boundary value $F(\lambda+i0)$ for a.e. $\lambda$ in $\mathbb{R}$. Moreover, by Theorem 11.5, if $\alpha$ is fixed, $F(\lambda+i0)\neq -\alpha^{-1}$ for a.e. $\alpha$. Thus we can define $\arg(1+\alpha F(\lambda+i0))$ for a.e. $\lambda$. Since $\operatorname{Im} F(\lambda+i0)\geq 0$, if $\alpha>0$ (resp. $\alpha<0$), this arg lies in $[0,\pi]$ (resp. $[-\pi,0]$). If $1+\alpha F(\lambda+i0)\in(-\infty,0)$, that sets the argument to $(\operatorname{sgn}\alpha)\pi$.

DEFINITION. The Krein spectral shift, $\xi_\alpha(\lambda)$, is defined by

$$\xi_\alpha(\lambda)=\frac{1}{\pi}\arg(1+\alpha F(\lambda+i0))$$

DEFINITION. $i(\alpha)\equiv\min(\inf\operatorname{spec}(A),\inf\operatorname{spec}(A_\alpha))$.

THEOREM 11.9. (i) *$0\leq\pm\xi_\alpha(\lambda)\leq 1$ if $\pm\alpha>0$; $\xi_\alpha(\lambda)=0$ if $\lambda<i(\alpha)$.*

(ii)

$$\operatorname{Tr}((A-z)^{-1}-(A_\alpha-z)^{-1})=\int(\lambda-z)^{-2}\xi_\alpha(\lambda)\,d\lambda \tag{11.19}$$

*if $z\in\mathbb{C}\backslash[i(\alpha),\infty)$.*

(iii) *If $\|\varphi\|<\infty$, then $\xi_\alpha\in L^1$ and*

$$\int\xi_\alpha(\lambda)\,d\lambda=\alpha\|\varphi\|^2 \tag{11.20}$$

PROOF. (i) The first part follows from the discussion above. If $\alpha>0$, $i(\alpha)=\inf\operatorname{spec}(A)$ and $F(\lambda+i0)>0$ for $\lambda<i(\alpha)$ so $\xi_\alpha(\lambda)=0$. If $\alpha<0$, $i(\alpha)$ is either $\inf\operatorname{spec}(A)$ or the unique $\lambda<i(\alpha)$ with $1+\alpha F(\lambda)=0$. Either way, since $F(\lambda)$ is monotone increasing on $(-\infty,i(0))$, below $i(\alpha)$, we have that $1+\alpha F(\lambda)>0$.

(ii) Let $H_\alpha(z)=\log(1+\alpha F(z))$ on $(-\infty,i(\alpha))$ analytically continued to $\mathbb{C}\backslash[i(\alpha),\infty)$. Consider a contour, $C$, which goes from $i(\alpha)$ just above the real axis to $R$, then directly up to $R+iR$, then on a circle of radius $R\sqrt{2}$ to $R-iR$, then up to just above the real axis, then along the bottom of the real axis to $i(\alpha)$. Consider for $z_0\in(-\infty,i(\alpha))$,

$$\frac{1}{2\pi i}\oint_C\frac{H_\alpha(z)}{(z-z_0)^2}\,dz\equiv I_\alpha(z_0)$$

By Cauchy's formula, once $R\sqrt{2}>|z_0|$, the integral is just $dH_\alpha/dz$, which is the left side of (11.19) by (11.16).

Since $|\operatorname{Im}H_\alpha|\leq\pi$ and the real parts of the contour from $R$ to $R+iR$ and from $R$ to $R-iR$ cancel, and since $F(z)\to 0$ as $z\to\infty$ with $|\arg z|>\pi/2$, the infinite part of the contour contributes 0 as $R\to\infty$. Since $|\operatorname{Im}H_\alpha|\leq\pi$, dominated convergence lets us take the contour to the real axis and get $\operatorname{Im}\log(1+\alpha F(z))=\arg(1+\alpha F(z))$. Thus, we get the right side of (11.19).

(iii) By Theorem 11.1(iii) and (iv) and the fact that $\int d\mu_0(z) = \|\varphi\|^2$, we see that

$$\lim_{\gamma\to\infty} \gamma^2 \frac{d}{dz} H_\alpha(-\gamma) = \alpha\|\varphi\|^2$$

so, by (ii),

$$\lim_{\gamma\to\infty} \int \frac{\gamma^2}{(x+\gamma)^2}\, \xi_\alpha(x)\, dx = \alpha\|\varphi\|^2$$

By monotone convergence, we obtain (11.20). □

The next result extends (11.19) to more general functions than $f(x) = (x-z)^{-1}$. The formula we want is:

$$\operatorname{Tr}(f(A) - f(A_\alpha)) = -\int f'(x)\xi_\alpha(x)\, dx \tag{11.21}$$

Formally, this holds for any reasonable $f$ since it holds for $f(x) = (x-z)^{-1}$ and those $f$'s are "dense" in reasonable $f$. One issue is that we don't even know that $f(A) - f(A_\alpha)$ is trace class! Our hypotheses on $f$ are not optimal, but include the critical example $f(x) = e^{-sx}$ (cutoff near $-\infty$).

THEOREM 11.10. *Let $f$ be a $C^2$ on $\mathbb{R}$ with $(1+|x|)^2 f^{(j)} \in L^2(0,\infty)$ for $j = 1, 2$. Then for any $\alpha$, $f(A) - f(A_\alpha)$ is trace class and* (11.21) *holds.*

PROOF. We only sketch the details, supposing without loss of generality that $\alpha > 0$ and $A \geq 1$ (otherwise, reverse the roles and/or add a constant to $A$). And we write $B \equiv A_\alpha$ for notational simplicity.

1. $B^{-1} - A^{-1}$ is rank one and so trace class. It follows that $B^{-2}e^{isB} - A^{-2}e^{isA}$ is trace class with trace norm bounded by $c(1+|s|)$ since

$$\begin{aligned} B^{-2}e^{isB} - A^{-2}e^{isA} &= B^{-1}e^{isB}(B^{-1} - A^{-1}) \\ &\quad - i\int_0^s e^{iuB}(B^{-1} - A^{-1})e^{i(s-u)A} + (B^{-1} - A^{-1})e^{isA}A^{-1} \end{aligned}$$

2. Let $\widehat{f}$ be the Fourier transform of $f$. Under the hypothesis,

$$\int (1+|k|)\left[\left|\frac{d\widehat{f}}{dk}\right| + \left|\frac{d^2\widehat{f}}{dk^2}\right|\right] dk < \infty \tag{11.22}$$

and that implies that we can integrate by parts twice to find that

$$f(B) - f(A) = -(2\pi)^{-1/2} \int \left[\frac{d^2\widehat{f}}{dk^2}(k)\right] [B^{-2}e^{ikB} - A^{-2}e^{ikA}]\, dk$$

It follows by Step 1 that $f(B) - f(A)$ is trace class.

3. From (11.19), one deduces (11.21) for $f(x) = x^{-2}e^{ixs}$ by writing

$$f(x) = \lim_{n\to\infty} \left[x^{-2}\left(1 - \frac{ixs}{n}\right)^{-n}\right]$$

for one can use analyticity to get (11.21) for $f(x) = (x-z)^{-n}$ and then for this special $f$.

4. (11.21) for general $f$ follows from the formula in Step 2 and the equality proven in Step 3. □

REMARK. This does not capture all $f$ for which (11.21) holds. For example, $f(x) = (x+1)^{-\beta}$ with $0 < \beta < 1$ does not obey the hypothesis, but using

$$(A+1)^{-\beta} = \frac{\sin \pi\beta}{\pi} \int_0^\infty (A+1+x)^{-1} |x|^{-\beta}\, dx$$

and $\|(A+x)^{-1} - (A_\alpha + x)^{-1}\|_1 \le C(|x|+1)^{-2}$ for $x$ large, one can prove (11.21) for such $f$.

PROPOSITION 11.11. *Suppose that $(a,b)$ is an interval disjoint from* $\mathrm{spec}(A) \cup \mathrm{spec}(A_\alpha)$. *Then*
(i) *$\xi_\alpha(x)$ is constant on $(a,b)$.*
(ii) *Its value is either* 0 *or* 1.
(iii) *$P_{(-\infty,a)}(A_\alpha) - P_{(-\infty,a)}(A)$ is trace class and its trace is the common value of $\xi_\alpha(x)$ in $(a,b)$.*

PROOF. (i) Let $f \in C_0^\infty(a,b)$. Then, $f(A) = f(A_\alpha) = 0$ so $\int f'(x)\xi_\alpha(x)\,dx = 0$ by Theorem 11.10. Thus the distributional derivative $d\xi_\alpha/dx = 0$ on $(a,b)$, so $\xi_\alpha$ is constant there.

(ii) It is a general fact about the trace of a difference of projections that is trace class that it is an integer [**108, 22**]. Since $0 \le \xi_\alpha(x) \le 1$, it must be 0 or 1. [There is a simple argument: Since $(a,b) \cap \mathrm{spec}(A) = \emptyset$, $F(x+i0)$ is real and continuous on $(a,b)$. If $F(x_0+i0) = -\alpha^{-1}$ anywhere on $(a,b)$, then $F_\alpha(z) = F(z)/[1+\alpha F(z)]$ has a pole at $x_0$, and so $x_0$ is an eigenvalue of $A_\alpha$. Since $(a,b) \cap \mathrm{spec}(A_\alpha) = \emptyset$, this cannot happen; that is, $1 + \alpha F(x+i0)$ has no zeros, and so its argument is constant at either 0 or 1.]

(iii) Let $f$ be a function in $C^\infty(\mathbb{R})$ which is 1 on $(i(\alpha), a)$, 0 on $[b,\infty)$. Then $f(A) = P_{(-\infty,a)}(A)$, $f(A_\alpha) = P_{(-\infty,a)}(A_\alpha)$, and $-\int f'(x)\xi_\alpha(x) = -\int_a^b f'(x)\xi_\alpha(x) = -\xi_\alpha(x_0)\int_a^b f'(x)\,dx = \xi_\alpha(x_0)$, the common value of $\xi_\alpha(x)$ in $(a,b)$. □

REMARK. Because of (iii), $\xi_\alpha(x)$ is called a spectral shift.

There is an interesting extension of Theorem 11.8, an abstraction of a result of Javrjan [**161**]:

THEOREM 11.12. *Let $\beta < \gamma$. Then*

$$\int_\beta^\gamma (d\mu_\alpha(E))\, d\alpha = (\xi_\gamma(E) - \xi_\beta(E))\, dE \tag{11.23}$$

REMARK. If $c \neq 0$, $\lim_{\alpha\to\infty} \arg(1+\alpha c) - \arg(1-\alpha c) = \pi$, so $\lim_{\alpha\to\infty}[\xi_\alpha(E) - \xi_{-\alpha}(E)] = 1$ and (11.23) implies (11.17).

PROOF. By (11.13), if $\mathrm{Im}\, z \neq 0$,

$$\begin{aligned}
\int_\beta^\gamma \left[\int \frac{d\mu_\alpha(E)}{(E-z)^2}\right] d\alpha &= \int_\beta^\gamma \frac{d}{dz}\left[\frac{F(z)}{1+\alpha F(z)}\right] d\alpha \\
&= \frac{d}{dz}\int_\beta^\gamma \frac{F(z)}{1+\alpha F(z)}\, d\alpha \\
&= \frac{d}{dz}\,[\log(1+\gamma F(z)) - \log(1+\beta F(z))] \\
&= \int \frac{\xi_\gamma(E) - \xi_\beta(E)}{(E-z)^2}\, dE
\end{aligned}$$

proving (11.23) smeared with $(E-z)^2$. Such functions have linear combinations which are dense in a big enough space to imply (11.23) as an equality of measures. □

Since $d\mu_\alpha$ obeys $\int d\mu_\alpha(\lambda)(|\lambda|+1)^{-1} < \infty$, with bounds uniform in $\alpha$ on compacts, this result implies a similar estimate for $\xi_\alpha$. Here is another proof:

THEOREM 11.13. (i) *If $\alpha < \infty$, then*

$$\int |\xi_\alpha(E)|(1+|E|)^{-1}\,dE < \infty \tag{11.24}$$

*and*

$$\int \xi_\alpha(E)(E-z)^{-1}\,dE = \log(1+\alpha F(z)) \tag{11.25}$$

(ii) $\int |\xi_\alpha(E)|\,dE \equiv \infty$ *if and only if* $\varphi \notin \mathcal{H}$.

PROOF. (i) Integrating

$$\frac{d}{dz}\log(1+\alpha F(z)) = \int \xi_\alpha(y)(y-z)^{-2}\,dy$$

from $-w$ to $-x$ for $x < w$ real, we get

$$\log(1+\alpha F(-x)) - \log(1+\alpha F(-w)) = \int \xi_\alpha(y)\frac{1}{y+x}\left(\frac{w-x}{y+w}\right)dy$$

As $w \to \infty$, $F(-w) \to 0$ so $\log(1+\alpha F(-w)) \to 0$ while $\frac{w-x}{y+w} = (1+\frac{y+x}{w-x})^{-1}$ is monotone increasing in $w$, so (11.24) is proven and (11.25) is proven for $z=-x$, $x$ real. By analytic continuation, it holds for all $z \in \mathbb{C}\backslash[i(\alpha),\infty)$.

(ii) By (11.25) and the monotone convergence theorem,

$$\begin{aligned}\int |\xi_\alpha(E)|\,dE &= (\operatorname{sgn}\alpha)\lim_{x\to\infty}\int \frac{x}{E+x}\,[\xi_\alpha(E)]\,dE \\ &= \operatorname{sgn}\alpha \lim_{x\to\infty} x\log(1+\alpha F(-x)) \\ &= |\alpha|\,\|\varphi\|^2\end{aligned}$$

by Theorem 11.1(iv). □

REMARK. It is not hard to show that $\int x^n|\xi_\alpha(x)|\,dx < \infty$ if and only if $\varphi \in \mathcal{H}_{n/2}(A)$, and in that case,

$$\int x^n \xi_\alpha(x)\,dx = \operatorname{Tr}\frac{A_\alpha^{n+1} - A^{n+1}}{n+1}$$

### Appendix to Section 11.4: The Krein Spectral Shift for Trace Class Perturbations

Let $C - A = (\varphi,\,\cdot\,)\varphi$ be rank one with $\|\varphi\| < \infty$ and $C \geq A$. Then for nice $f$, we have

$$\operatorname{Tr}(f(C)-f(A)) = \int f'(x)\xi_{C,A}(x)\,dx \tag{11.26a}$$

where

$$\int \xi_{C,A}(x)\,dx = \operatorname{Tr}(C-A) \tag{11.26b}$$

and

$$\int |\xi_{C,A}(x)|\, dx = \mathrm{Tr}(|C-A|) \tag{11.26c}$$

If $C-A$ is finite rank, we can diagonalize it and get it as a sum of rank one perturbations, and (11.26a)–(11.26c) continue to hold. By (11.26c), $\xi$ has a limit in $L^1$ norm as $C-A$ approaches a general trace class of operator. This proves

THEOREM 11.14 (Krein's Spectral Shift Formula). *If $A$ is selfadjoint and bounded below, $B$ is trace class, and $C = A+B$, then there exists an $L^1$ function, $\xi_{C,A}(x)$, on $\mathbb{R}$ vanishing for $x$ near $-\infty$ so that* (11.26a)–(11.26c) *hold.*

REMARK. $\xi$ is no longer given as $\arg(1+\alpha F(z))$ with $F(z)$ as simple as in the rank one case. However, in the rank one case,

$$\begin{aligned}\mathrm{Tr}(1+\alpha F(z)) &= 1 + \mathrm{Tr}((C-A)(A-z)^{-1}) \\ &= \det(1+(C-A)(A-z)^{-1}) \\ &= \det((C-z)(A-z)^{-1})\end{aligned}$$

and the argument of this has the correct additive properties. The result is [**178**]

$$\xi_{C,A}(\lambda) = \lim_{\epsilon\downarrow 0} \arg[\det(1+(C-A)(A-\lambda-i\epsilon)^{-1})]$$

where det is the determinant as defined in Chapter 3.

## 11.5. Infinite Coupling

In the rank one setting we have been describing, let $P$ be the projection onto the space spanned by $\varphi$ if $\|\varphi\| < \infty$ and $P = 0$ if $\|\varphi\| = \infty$.

Look at the formula (11.15) for $(A_\alpha - z)^{-1}$. It clearly shows the first part of

THEOREM 11.15. (i) *As $\alpha \to \pm\infty$, $(A_\alpha - z)^{-1}$ converges in norm to a limit we call $(A_\infty - z)^{-1}$ given by*

$$(A_\infty - z)^{-1} = (A-z)^{-1} - F(z)^{-1}((A-\bar{z})^{-1}\varphi, \cdot\,)(A-z)^{-1}\varphi \tag{11.27}$$

(ii) *If $\|\varphi\| = \infty$, then there is a selfadjoint operator $A_\infty$ on $\mathcal{H}$ whose resolvent is $(A_\infty - z)^{-1}$. If $\|\varphi\| < \infty$, then there is a selfadjoint operator, $\tilde{A}_\infty$, on $(1-P)\mathcal{H}$ so that $(A_\infty - z)^{-1}\eta = (\tilde{A}_\infty - z)^{-1}(1-P)\eta$. (We'll call $\tilde{A}_\infty$ as just $A_\infty$ with the understanding that $(A_\infty - z)^{-1} = 0$ on $P\mathcal{H}$ and we'll write $\mathcal{H}(A_\infty) = (1-P)\mathcal{H}$.)*

(iii) *$\mathcal{H}_{+1}(A_\infty) = \{\psi \in \mathcal{H}_{+1}(A) \mid (\varphi,\psi) = 0\}$ and*

$$(\psi, A_\infty\psi) = (\psi, A\psi)$$

*for $\psi \in \mathcal{H}_{+1}(A_\infty)$.*

PROOF. We have proven (i). To prove (ii), (iii), one uses the following monotone convergence theorem for forms ([**164, 301**]):

Let $C_n$ be a family of nonnegative selfadjoint operators with $\mathcal{H}_{+1}(C_{n+1}) \subseteq \mathcal{H}_{+1}(C_n)$ and $(\psi, C_{n+1}\psi) \geq (\psi, C_n\psi)$ if $\psi \in \mathcal{H}_{+1}(C_{n+1})$. Let

$$\mathcal{H}_{+1}(C_\infty) = \left\{\psi \in \bigcap_n \mathcal{H}_{+1}(C_n) \;\middle|\; \sup_n (\psi, C_n\psi) < \infty\right\}$$

Let $c_\infty$ be the quadratic form on $\mathcal{H}_{+1}(C_\infty)$ given by $c_\infty(\psi,\psi) = \sup_n(\psi, c_n\psi)$. Then, there is a selfadjoint operator $C_\infty$ on $\overline{\mathcal{H}_{+1}(C_\infty)}$ so that if $(C_\infty - z)^{-1}$ is extended to $\mathcal{H}_{+1}(C_\infty)^\perp$ by setting it equal to zero there, then $(C_n - z)^{-1}$ converges strongly to $(C_\infty - z)^{-1}$.

This yields (ii), (iii) if one notes that

$$\sup_\alpha(\psi, A_\alpha\psi) < \infty \Leftrightarrow (\psi, \varphi) = 0 \qquad \square$$

Since $\xi_\alpha(x)$ is monotone in $\alpha$ and bounded by 1, it has a limit $\xi_\infty(x)$. In fact,

$$\xi_\infty(x) = \arg(F(x + i0)) \qquad 0 \leq \xi_\infty(x) \leq 1 \tag{11.28}$$

since

$$\xi_\alpha(x) = \arg(\alpha^{-1} + F(x + i0))$$

By a limiting argument, $\xi_\infty(x)\,dx = \int_0^\infty (d\mu_\alpha(x))\,d\alpha$.

In addition,

$$\operatorname{Tr}((A - z)^{-1} - (A_\infty - z)^{-1}) = \lim_{\alpha\to\infty} \frac{\alpha F'(z)}{1 + \alpha F(z)} = \frac{F'(z)}{F(z)} \tag{11.29a}$$

$$= \frac{d}{dz} \log F(z) \tag{11.29b}$$

so by mimicking the arguments in the $\alpha < \infty$ case,

$$\operatorname{Tr}(f(A) - f(A_\infty)) = -\int f'(x)\xi_\infty(x)\,dx$$

for the $f$'s in Theorem 11.10. One place that $\xi_\alpha$ and $\xi_\infty$ differ though is (compare with (11.24))

PROPOSITION 11.16 (Gesztesy-Simon [**131**]).

$$\int (1 + |x|)^{-1}\xi_\infty(x)\,dx = \infty$$

PROOF. By monotone convergence,

$$\int (1 + x)^{-1}\xi_\infty(x)\,dx = \lim_{\alpha\to\infty} \int (1 + x)^{-1}\xi_\alpha(x)\,dx$$

$$= \lim_{\alpha\to\infty} \log(1 + \alpha F(-1)) = \infty \qquad \square$$

If $A$ is bounded, it is easy to see why the integral diverges; $\xi_\infty(x) = 1$ for $x > \|A\|$, so the integral diverges logarithmically exactly as $\log x$. This is typical of the case when $\|\varphi\| < \infty$.

THEOREM 11.17 (Gesztesy-Simon [**131**]). *If $\|\varphi\| < \infty$, then $\int (1 + |x|)^{-1}[1 - \xi_\infty(x)]\,dx < \infty$ and*

$$\log\left[\frac{(-z)F(z)}{\|\varphi\|^2}\right] = \int_0^\infty \frac{[\xi_\infty(x) - 1]}{x - z}\,dx \tag{11.30}$$

PROOF. We have that

$$\frac{d}{dz} \log F(z) = \int \frac{\xi_\infty(x)}{(x - z)^2}\,dx \tag{11.31}$$

while clearly,

$$\frac{d}{dz} \log(-z^{-1}) = \int_0^\infty \frac{1}{(x - z)^2}\,dx \tag{11.32}$$

since for $z < 0$, both sides are just $(-z)^{-1}$. (11.32) is a direct calculation but can also be regarded as (11.31) for the case $F(z) = -1/z$.

By (11.31)–(11.32),

$$\frac{d}{dz}\log\left(\frac{(-z)F(z)}{\|\varphi\|^2}\right)=\int_0^\infty \frac{(\xi_\infty(x)-1)}{(x-z)^2}\,dx$$

By Theorem 11.1(iv), $\log((-z)F(z)/\|\varphi\|^2)\to 0$ as $z\to-\infty$ along the real axis. The same monotone convergence theorem argument that we used in the proof of Theorem 11.13(i) and the fact that $1-\xi_\alpha(x)\geq 0$ implies that $(1+|x|)^{-1}(1-\xi_\infty)$ is in $L^1$ and that (11.30) holds. □

In Section 14.1, we will extend this idea to some cases where $\|\varphi\|=\infty$.

How can one get a handle on a spectral measure for $A_\infty$? At first sight, it seems a hard job since

$$\lim_{\alpha\to\infty}\int\frac{d\mu_\alpha(x)}{x-z}=\lim_{\alpha\to\infty}\frac{F(z)}{1+\alpha F(z)}=0$$

so $d\mu_\alpha(x)\to 0$ weakly. The key is to renormalize! Define

$$d\rho_\alpha(x)=(1+\alpha^2)\,d\mu_\alpha(x)$$

One can prove the following:

THEOREM 11.18 (Gesztesy-Simon [**131**]). (i) *There exists a measure $d\rho_\infty$ (which is nonzero except for the trivial case where $\varphi$ is an eigenvector of $A$) so that*

$$\int\frac{d\rho_\infty(x)}{(1+|x|)^2}<\infty$$

*and*

$$\lim_{\alpha\to\infty}\int f(x)\,d\rho_\alpha(x)=\int f(x)\,d\rho_\infty(x)\tag{11.33}$$

*for any $f\in C_0^\infty(\mathbb{R})$.*

(ii) *There exists a vector $\eta\in\mathcal{H}_{-2}(A_\infty)$ so that $d\rho_\infty$ is the spectral measure $d\mu^\eta_{A_\infty}$. If $\varphi$ is cyclic for $A$, then $\eta$ is cyclic for $A_\infty$.*

(iii) *(11.33) holds for $f(x)=(x-z)^{-3}$, but it only holds for $f(x)=(x-z)^{-2}$ if $\|\varphi\|=\infty$. Explicitly,*

$$\lim_{\alpha\to\infty}\int\frac{d\rho_\alpha(x)}{(x-z)^2}=\|\varphi\|^{-2}+\int\frac{d\rho_\infty(x)}{(x-z)^2}$$

(iv) *$d\rho_\infty$ is the boundary value of $-F(z)^{-1}$ in the sense that for $f\in C_0^\infty(\mathbb{R})$,*

$$\int f(x)\,d\rho_\infty(x)=\lim_{\epsilon\downarrow 0}\frac{1}{\pi}\int f(x)\,\mathrm{Im}[-F(x+i\epsilon)^{-1}]\,dx$$

We refer the reader to [**131**] for a complete proof, but will make a series of remarks that illuminate the result and partly prove it.

REMARKS. 1. Notice that $\eta$ is only stated to be in $\mathcal{H}_{-2}(A_\infty)$, not $\mathcal{H}_{-1}(A_\infty)$. Indeed, the differential equation case of the next section when $V=0$ is one where $d\rho_\infty(x)\sim cx^{1/2}\,dx$ and $\eta\notin\mathcal{H}_{-1}(A_\infty)$. Indeed [**131**], for any $\theta<2$, there are examples where $\int d\rho_\infty(x)/(1+|x|)^\theta=\infty$.

2. If $A$ is bounded, $\eta=(1-P)A\varphi$. Otherwise, one has that $\eta$ obeys

$$(A_\infty-z)^{-1}\eta=(1-P)F(z)^{-1}(A-z)^{-1}\varphi$$

(with the convention that $P=0$ if $\|\varphi\|=\infty$).

3.
$$\int \frac{d\mu_\alpha(x)}{(x-z)^2} = \frac{d}{dz}\,\frac{F(z)}{1+\alpha F(z)} = \frac{F'(z)}{(1+\alpha F(z))^2}$$
so
$$\lim_{\alpha\to\infty} \int \frac{d\rho_\alpha(x)}{(x-z)^2} = \frac{F'}{F^2} \tag{11.34}$$
so the $\|\varphi\|^{-2}$ term in (iii) is related to Theorem 11.1(v).

4. Notice that the right side of (11.34) is $\frac{d}{dz}(-\frac{1}{F})$. This is the key to proving (iv).

There is an interesting reparametrization of the $\rho$'s that is exactly what is used in the differential equation case. For $\theta \in (0,\pi)$, we can define $\alpha = -\cot(\theta)$ so $\alpha$ runs from $-\infty$ to $\infty$ as $\theta$ runs from 0 to $\pi$. Let
$$d\tilde\rho_\theta(x) = d\rho_{-\cot(\theta)}(x) = [\sin^2(\theta)]^{-1}\, d\mu_{-\cot(\theta)}(x)$$
Then, $d\tilde\rho_\theta$ has a limit as $\theta \to 0, \pi$ which is just what we called $d\rho_\infty(x)$. Moreover, we claim that
$$\int_\theta^\pi (d\tilde\rho_\theta(E))\, d\theta = dE \tag{11.35}$$
for
$$\int_0^\pi d\tilde\rho_\theta(E)\, d\theta = \int_{-\infty}^\infty (1+\alpha^2)\, d\mu_\alpha(E)\, \frac{d\alpha}{1+\alpha^2} = dE$$

## 11.6. Boundary Condition Variation of ODE's

For applications to Schrödinger operators, there are two main categories of the general theory: variation of potential in the discrete (Jacobi) case, which we'll discuss in the next section, and the subject of this section, variation of boundary condition.

We begin with a lightning review of the standard Weyl $m$-function approach (see [**330**]) and then rephrase the setup into the context of these lectures. We will consider a Schrödinger operator on $(0,\infty)$ (and later on $(-\infty,\infty)$) of the form
$$H_\theta u = -u''(x) + V(x)\,u(x) \tag{11.36}$$
where, for simplicity, we suppose that $V(x)$ is bounded from below but not necessarily continuous. This makes $H$ limit point at infinity so we only need a boundary condition at zero, which is where the parameter $\theta$ enters. The boundary condition at 0 is
$$u(0)\cos\theta + u'(0)\sin\theta = 0 \qquad 0 \le \theta < \pi \tag{11.37}$$
so that $\theta = 0$ is the Dirichlet boundary condition and $\theta = \pi/2$ is the Neumann boundary condition.

Explicitly, $H_\theta$ defines a selfadjoint operator with domain
$$D(H_\theta) = \{u \in L^2(0,\infty)\ |u, u' \text{ are locally a.c. functions,}$$
$$u \text{ obeys (11.37) and } H_\theta u \in L^2\}$$

We will often be interested in solutions of the differential equation
$$-u'' + Vu = zu \tag{11.38}$$
and will consider the Wronskian $W(f,g)$ of two functions
$$W(f,g)(x) = f(x)g'(x) - f'(x)g(x)$$

As usual, if $f, g$ solve (11.38) for the same $z$, then $W(f,g)(x)$ is independent of $x$.

Given $\theta$, three solutions of (11.38) will concern us: $\phi$ which obeys the boundary condition (11.37), $\eta$ defined with a "dual" boundary condition, and $\psi_+$ for $\operatorname{Im} z \neq 0$, the solution which is $L^2$ near $x = \infty$. Explicitly, $\phi, \eta$ are defined by boundary values at $x = 0$:

$$\phi_\theta(z,0) = -\sin(\theta) \qquad \phi'_\theta(z,0) = \cos(\theta)$$
$$\eta_\theta(z,0) = \cos\theta \qquad \eta'_\theta(z,0) = \sin(\theta)$$

so

$$W(\eta_\theta, \phi_\theta)(x) = 1 \tag{11.39}$$

The third solution, $\psi_+$, is defined with mixed boundary conditions; it is in $L^2((0,\infty))$, which determines it up to a multiplicative constant, and it is normalized by

$$W(\psi_{+,\theta}, \phi_\theta) = \psi_{+,\theta}(0)\cos\theta + \psi'_{+,\theta}(0)\sin\theta = 1 \tag{11.40}$$

Since $\phi, \eta$ are independent solutions of (11.38), we have $\psi_+ = a\eta + b\phi$. By (11.39) and (11.40), $a = 1$. We therefore define the Weyl $m$-function, $m_\theta(z)$, by

$$\psi_{+,\theta}(z,x) = \eta_\theta(z,x) + m_\theta(z)\phi_\theta(z,x) \tag{11.41a}$$

or equivalently,

$$m_\theta(z) = W(\eta_\theta, \psi_{+,\theta}) \tag{11.41b}$$

or

$$m_\theta(z) = -\lim_{R\to\infty} \frac{\eta_\theta(z,R)}{\phi_\theta(z,R)} \tag{11.41c}$$

The last (which follows from $\psi_+ \to 0$ at infinity) is the usual definition in the limit point case because it is what Weyl used, although the first two are more useful.

It can be seen that

$$\operatorname{Im} z > 0 \Rightarrow \operatorname{Im} m_\theta(z) > 0 \tag{11.42}$$

We won't prove this now, since it will follow from the general theory of Borel transforms once we have set up the connection. The spectral measure $d\rho_\theta$ is defined by

$$d\rho_\theta(x) = \lim_{\epsilon\downarrow 0} \frac{1}{\pi} [\operatorname{Im} m_\theta(x+i\epsilon)]\, dx$$

For $\theta \neq 0$, as we'll see,

$$m_\theta(z) = \cot(\theta) + \int \frac{d\rho_\theta(x)}{x-z} \tag{11.43}$$

The Green's function, $G_\theta(x,x';z)$, is the integral kernel of $(H_\theta - z)^{-1}$, that is,

$$((H_\theta - z)^{-1}u)(x) = \int G_\theta(x,x';z)u(x')\, dx'$$

Because of the normalization (11.40), one has

$$G_\theta(x,x';z) = \phi_\theta(z,x_<)\psi_{+,\theta}(z,x_>)$$

where $x_< = \min(x,x')$, $x_> = \max(x,x')$. In particular, by (11.41a),

$$G_\theta(0,0;z) = +\sin^2\theta(-\cot\theta + m_\theta(z)) \tag{11.44}$$

As a final general formula, we note that Wronskian gymnastics (we'll get it other ways below) shows that

$$m_\kappa(z) = \frac{-\sin(\kappa-\theta) + \cos(\kappa-\theta)m_\theta(z)}{\cos(\kappa-\theta) + \sin(\kappa-\theta)m_\theta(z)} \tag{11.45a}$$

In particular,

$$m_0(z) = -\frac{1}{m_{\pi/2}(z)} \tag{11.45b}$$

We want to show how to think of this in terms of rank one perturbations. Let $u \in D(H_\theta)$ be real-valued. Then, integrating by parts,

$$\begin{aligned}(u, H_\theta u) &= \int_0^\infty (u(x))[-u'' + Vu](x)\,dx \\ &= \int_0^\infty [u'(x)^2 + V(x)u^2(x)]\,dx + u'(0)u(0) \\ &\equiv \tau(u,u) - \cot(\theta)u(0)^2\end{aligned} \tag{11.46}$$

because of the boundary condition.

Let $A$ be the $\theta = \pi/2$ Neumann boundary condition operator. A simple Sobolev estimate shows that for any $u \in \mathcal{H}_{+1}(A)$ is continuous in $x$ and

$$|u(0)|^2 \le c(u, (A+1)u)$$

that is, $\delta(x)$, the Dirac delta function, lies in $\mathcal{H}_{-1}(A)$. We can form

$$A_\alpha = A + \alpha\delta(x)$$

(11.46) says that $H_\theta = A_{-\cot(\theta)}$. Moreover,

$$F_\alpha(z) = (\delta, (A_\alpha - z)^{-1}\delta)$$

is exactly $G_\theta(0,0;z)$. Since $G$ is related to $m$ by (11.44), (11.45a)/(11.45b) is just the relation $F_\alpha(z) = F(z)/[1+\alpha F(z)]$.

In addition, because of (11.44) and (11.43), we have

$$d\rho_\theta(E) = \sin^2\theta\, d\mu_{-\cot(\theta)}(E)$$

that is, the spectral measure $d\rho$ is precisely the measure we called $d\tilde\rho_\theta$ in the last section. The Dirichlet spectral measure, $d\tilde\rho_0$, is an example of a $d\rho_\infty$ with $\int d\rho_\alpha(E)/[|E|+1] = \infty$. (11.45b) is just Theorem 11.18(iv). The vector $\eta$ of Section 11.5 is just $\delta'$; see [**131**].

The point of this section is the spectral analysis of rank one perturbations in the next chapter will apply to boundary condition variation directly (rather than by analogy).

## 11.7. Jacobi Matrices

Jacobi matrices are operators on $\ell^2(\mathbb{Z}^\nu)$ which are discrete analogs of Schrödinger operators. We write $u \in \ell^2(\mathbb{Z}^\nu)$ as a function on $\mathbb{Z}^\nu$. Given a function, $V$, on $\mathbb{Z}^\nu$, we define the associated Jacobi matrix, $h$, by

$$(hu)(n) = \sum_{|j|=1} u(n+j) + V(n)u(n) \tag{11.47a}$$

which we write as

$$h = h_0 + V \tag{11.47b}$$

Let $\delta_0$ be the vector in $\ell^2$ which is given by

$$\delta_0(n) = \delta_{n0}$$

Let $P$ be the projection on $\delta_0 : (\delta_0, \cdot)\delta_0$. Fix $V$ and let $A$ be the Jacobi matrix associated to $V - \delta_{\cdot 0}(V(0))$. Then

$$h = A + V(0)P$$

and variations of $V$ are precisely rank one perturbations.

As an application of this idea, let $V_\omega(n)$ be a random process where the $V_\omega(n)$ are independent, identically distributed random variables with distribution $p(v)\,dv$ where $p$ is bounded. There is a fundamental quality called the integrated density of states, $k(\lambda)$, which is given by the following: Let $d\mu_\omega^{(0)}$ be the spectral measure for $h_0 + V_\omega$ with vector $\delta_0$. Then

$$k(\lambda) = E_\omega(\mu_\omega^{(0)}(-\infty, \lambda)) \tag{11.48}$$

This is not the basic definition but a derived formula for $k$ (see, e.g., [**23, 81**]). We claim

THEOREM 11.19 (Wegner's Estimate [**337**]).

$$|k(\lambda) - k(\lambda')| \leq c|\lambda - \lambda'|$$

*where* $c = \sup_\gamma p(\gamma)$.

PROOF. The result is equivalent to saying that $dk$ is absolutely continuous with respect to $d\lambda$ with $dk = q(\lambda)\,d\lambda$ with $\|q\|_\infty \leq c$. Fix $(V_\omega(n))_{n\neq 0}$. Then, by Theorem 11.8,

$$\int [d\mu_\omega^{(0)}(\lambda)]p(v(0))\,dv(0) \leq c \int d\mu_\omega^{(0)}(\lambda)\,dv(0) = c\,d\lambda$$

Averaging over $(V_\omega(n))_{n\neq 0}$, we get

$$E_\omega(d\mu_\omega) \leq c\,d\lambda$$

which is exactly what we want. □

CHAPTER 12

# Spectral Theory of Rank One Perturbations

We begin our analysis with the study of the absolutely continuous spectrum, obtaining a result that also follows from the trace class theory of scattering, but without wave operators. We next discuss the fundamental work of Aronszajn [**19**] and Donoghue [**20**], which identifies spectral information for $A_\alpha$ in terms of $F(z)$. Next, we discuss the Simon-Wolff [**314**] work on localization and apply it to one-dimensional random Jacobi matrices. Finally, we discuss recent results of Gordon [**138**] and del Rio et al. [**92**] on singular spectrum.

*Henceforth, we assume that* $\varphi$ *is a cyclic vector for* $A$*, that is,* $\{(A-z)^{-1}\varphi \mid z \in \mathbb{C}\}$ *is a total set for* $\mathcal{H}$. In general, if $\mathcal{H}_0$ is the closed subspace generated by these vectors, then $\mathcal{H}_0^\perp$ is an invariant space for each $A_\alpha$ and $A_\alpha = A$ on $\mathcal{H}_0^\perp$. Thus, the extension from the cyclic to general case is trivial. We make the assumption to be able to focus on the real issues and not have to state the theorems in convoluted forms to give the general case.

## 12.1. Invariance of the Absolutely Continuous Spectrum

Let $f(x)\,dx$ and $g(x)\,dx$ be two absolutely continuous (positive) measures on $\mathbb{R}$. Recall these measures are equivalent if and only if $\{x \mid f(x) \neq 0\}$ and $\{x \mid g(x) \neq 0\}$ agree up to sets of Lebesgue measure zero. Moreover, if $A$ is multiplication by $x$ on $L^2(\mathbb{R}, f\,dx)$ and $C$ is multiplication by $x$ on $L^2(\mathbb{R}, g\,dx)$, then $A$ and $C$ are unitarily equivalent if and only if $f\,dx$ and $g\,dx$ are equivalent. With these preliminaries:

THEOREM 12.1. *For all* $\alpha \neq \beta$*,* $(A_\alpha)_{\rm ac}$ *and* $(A_\beta)_{\rm ac}$*, the absolutely continuous parts of* $A_\alpha$ *and* $A_\beta$*, are unitarily equivalent.*

PROOF. Since $A_\alpha = A_\beta + (\alpha - \beta)(\varphi, \cdot\,)\varphi$, we can suppose that $\beta = 0$ without loss of generality. By the above discussion and Theorem 11.6(iv), it suffices to show that $S_1 \equiv \{E \mid \operatorname{Im} F(E+i0) \neq 0\}$ and $S_2 \equiv \{E \mid \operatorname{Im} F_\alpha(E+i0) \neq 0\}$ agree up to sets of measure zero. By the Aronszajn-Krein formula $F_\alpha = F/(1+\alpha F)$, which implies that

$$\operatorname{Im} F_\alpha(z) = \frac{\operatorname{Im} F(z)}{|1+\alpha F(z)|^2} \tag{12.1}$$

It follows that $S_1$ and $S_2$ agree up to the sets where $F(E+i0)$ fails to exist, the set where $F(E+i0) = \infty$, and the set where $F(E+i0) = -\alpha^{-1}$. These sets all have measure zero, so $S_1$ and $S_2$ agree up to sets of measure zero. $\square$

REMARKS. 1. The result extends to the noncyclic case and so extends to the finite rank case. It should be possible to accommodate the trace class case. This is an area under present study.

2. Since $(A_\infty + 1)^{-1} - (A_0 + 1)^{-1}$ is rank one in the usual sense, we can apply the theorem to $(A_\infty - 1)^{-1}$ and $(A_0 + 1)^{-1}$ and so, by a spectral mapping theorem,

extend the result to allow $\alpha = \infty$. One can also get $\alpha = \infty$ by using $d\rho_\infty$ and $\operatorname{Im}(-1/F) = \operatorname{Im} F/|F|^2$.

## 12.2. The Aronszajn-Donoghue Theory

Recall that

$$G(x) = \int \frac{d\mu(y)}{(x-y)^2}$$

The following basic theorem was proven by Aronszajn [**19**] for boundary condition dependence of spectrum for Sturm-Liouville operators and then extended to (bounded) rank one perturbations by Donoghue [**102**].

THEOREM 12.2. *For $\alpha \neq 0$ ($\alpha =$ infinity allowed with $\infty^{-1} = 0$), define*

$$S_\alpha = \{x \in \mathbb{R} \mid F(x+i0) = -\alpha^{-1};\, G(x) = \infty\}$$
$$P_\alpha = \{x \in \mathbb{R} \mid F(x+i0) = -\alpha^{-1};\, G(x) < \infty\}$$
$$L = \{x \in \mathbb{R} \mid \operatorname{Im} F(x+i0) \neq 0\}$$

*Then*

(i) $\{S_\alpha\}_{\alpha\neq 0; |\alpha|\le\infty}$, $\{P_\alpha\}_{\alpha\neq 0;|\alpha|\le\infty}$, *and $L$ are mutually disjoint.*
(ii) *$P_\alpha$ is the set of eigenvalues of $A_\alpha$. In fact,*

$$(d\mu_\alpha)_{\text{pp}}(x) = \sum_{x_n\in P_\alpha} \frac{1}{\alpha^2 G(x_n)}\,\delta(x-x_n) \qquad \alpha<\infty$$

$$(d\rho_\infty)_{\text{pp}}(x) = \sum_{x_n\in P_\infty} \frac{1}{G(x_n)}\,\delta(x-x_n) \qquad \alpha=\infty$$

(iii) *$(d\mu_\alpha)_{\text{ac}}$ is supported on $L$, $(d\mu_\alpha)_{\text{sc}}$ is supported on $S_\alpha$.*
(iv) *For $\alpha \neq \beta$, $(d\mu_\alpha)_{\text{sing}}$ and $(d\mu_\beta)_{\text{sing}}$ are mutually singular.*

REMARK. We say that a set $S$ "supports" a measure $d\nu$ if $\nu(\mathbb{R}\backslash S) = 0$.

PROOF. (i) is trivial and (iv) follows from (i), given (ii) and (iii). That $L$ supports $(d\mu_\alpha)_{\text{ac}}$ is proven in the last section.

By Theorem 11.6(iv), $(d\mu_\alpha)_{\text{sing}}$ is supported by $\{E \mid \lim_{\epsilon\downarrow 0}|F_\alpha(E+i\epsilon)| = \infty\}$. But

$$F_\alpha(z) = \frac{F(z)}{1+\alpha F(z)}$$

says that $|F_\alpha(E+i\epsilon)| \to \infty$ if and only if $F(E+i\epsilon) \to -1/\alpha$.

(ii), (iii) thus follow by Theorem 11.6(iii) and the claim that when $F(E+i\epsilon) \to -1/\alpha$ (even if $G(E)=\infty$),

$$\lim_{\epsilon\downarrow 0} \epsilon \operatorname{Im} F_\alpha(E+i\epsilon) = \frac{1}{[\alpha^2 G(E)]} \qquad (0<\alpha<\infty) \tag{12.2a}$$

$$\lim_{\epsilon\downarrow 0} \epsilon \operatorname{Im}(-F(E+i\epsilon))^{-1} = \frac{1}{G(E)} \qquad (\alpha=\infty) \tag{12.2b}$$

If $G(E)<\infty$, (12.2) follows from Theorem 11.2(v) which implies that $F(E+i\epsilon) = -1/\alpha + i\epsilon G(x) + o(\epsilon)$ and the formula for $F_\alpha$. If $G(E)=\infty$ and $\alpha=\infty$, Theorem 11.2(iv) implies $\epsilon^{-1}|F(x+i\epsilon)| \to \infty$, so $|\epsilon/F(x+i\epsilon)| \to 0$, proving (12.2b) in that case. If $G(E)=\infty$ and $0<\alpha<\infty$, by Theorem 11.2(iv) again, $\epsilon^{-1}\operatorname{Im}(1+\alpha F) \to \infty$, so $\epsilon|1+\alpha F|^{-1} \to 0$. Thus, since $F \to -1/\alpha$, $\epsilon|F/1+\alpha F| \to 0$. $\square$

This theorem has interesting consequences for one-dimensional Schrödinger operators and Jacobi matrices:

THEOREM 12.3. *Let $A$ be a Schrödinger operator on $[0,\infty)$ with Neumann boundary conditions at $x=0$ as in Section 11.6. Then $G(E)<\infty$ if and only if both of the following hold:*
(i) *$E$ is not an eigenvalue of $A$.*
(ii) *(11.38) (with $E=z$) has a solution which is $L^2$ at $+\infty$.*

PROOF. Since $G(E)$ finite implies that $F(E+i0)$ exists and is real and finite, Theorem 12.2 implies that $G(E)<\infty$ if and only if $E$ is an eigenvalue of $A+\alpha\delta$ for some $\alpha\neq 0$ in $\mathbb{R}\cup\{\infty\}$. But this is equivalent to (i), (ii). □

THEOREM 12.4. *Let $A$ be a Jacobi matrix on $\ell^2(\mathbb{Z})$ and $\varphi=\delta_0$ as in Section 11.7. Then $G(E)<\infty$ if and only if*
(i) *$E$ is not an eigenvalue of $A$.*
(ii) *One of the following holds:*
(a) *The equation*

$$u(n+1)+u(n-1)+V(n)u(n)=Eu(n) \tag{12.3}$$

*has an $\ell^2$ solution on $(0,\infty)$ with $u(0)=0$.*
(b) *(12.3) has an $\ell^2$ solution on $(-\infty,0)$ with $u(0)=0$.*
(c) *(12.3) has $\ell^2$ solutions $u_\pm$ on both $(0,\infty)$ and $(-\infty,0)$ with both $u_+(0)\neq 0$ and $u_-(0)\neq 0$.*

REMARKS. 1. To say $u$ is a solution on $(0,\infty)$ means $u$ is defined on $[0,\infty)$ but (12.3) holds for $n=1,2,\dots$.

2. In particular, if (12.3) has $\ell^2$ solutions on both $(0,\infty)$ and $(-\infty,0)$, then either $E$ is an eigenvalue of $A$ or $G(E)<\infty$.

PROOF. (a), (b) of (ii) hold if and only if $E$ is an eigenvalue of $A_\infty$. (c) holds if and only if $E$ is an eigenvalue of some $A_\alpha$ with $|\alpha|<\infty$. Given these, the proof is the same as for the last two theorems. □

## 12.3. The Simon-Wolff Criterion

The following is useful in discussing localization for random Jacobi matrices:

THEOREM 12.5 (Simon-Wolff [**314**]). *Fix an interval $[a,b]$. Then the following are equivalent:*
(a) *$G(E)<\infty$ for a.e. $E$ in $[a,b]$ (w.r.t. Lebesgue measure).*
(b) *For a.e. $\alpha$ (w.r.t. Lebesgue measure), $A_\alpha$ has only pure point spectrum in $[a,b]$.*

PROOF. Let $S=\{E\in[a,b]\mid G(E)=\infty\}$. Suppose (a) holds. Then $|S|=0$ by hypothesis. By Theorem 11.8, that means $\int\mu_\alpha(S)\,d\alpha=0$ so $\mu_\alpha(S)=0$ for a.e. $\alpha$. But if $\mu_\alpha(S)=0$, then in the language of Theorem 12.2, $\mu_\alpha(L\cap[a,b])=\mu_\alpha(S_\alpha\cap[a,b])=0$. It follows by that theorem that $\mu_\alpha$ has only point spectrum, so (b) holds.

Conversely, if $A_\alpha$ has only point spectrum on $[a,b]$, then $\mu_\alpha(S)=0$. If (b) holds, $\int\mu_\alpha(S)\,d\alpha=0$, so by Theorem 11.8, $|S|=0$ and (a) holds. □

As an example of how to apply this theorem, we want to prove localization for the Anderson model in one dimension following [**306**]. We'll use various facts about that model proven elsewhere.

THEOREM 12.6. *Let $\{v_\omega(x)\}_{n\in\mathbb{Z}}$ be bounded i.i.d.'s with distribution $p(\lambda)\,d\lambda$. Let $h_\omega = h_0 + v_\omega$ as in Section 11.7. Then for a.e. $\omega$, $h_\omega$ has only point spectrum.*

REMARKS. 1. The general proof following the steps below requires a lot less than bounded or independent random variables; see [**306**].

2. The proofs show that the eigenfunctions decay exponentially.

3. If $\operatorname{supp} p = [a,b]$, then $\operatorname{spec}(h_\omega) = [a-2, b+2]$ with probability one, so this is point spectrum dense in an interval.

PROOF. 1. The transfer matrix, $A_n(E,\omega)$, is the $2\times 2$ matrix defined by $\binom{u(n+1)}{u(n)} = A_n(E,\omega)\binom{u(1)}{u(0)}$ where $u$ obeys (12.3) for $v_\omega$. General principles (the subadditive ergodic theorem; see [**306**]) imply that for each fixed $E$,

$$\lim_{|n|\to\infty} \frac{1}{|n|} \log \|A_n(\omega,E)\| = \gamma(E) \tag{12.4}$$

for a.e. $\omega$ and a fixed $\gamma$. Obviously, whether the limit in (12.4) holds only depends on $v_\omega$ in a neighborhood of infinity.

2. It can be proven that $\gamma(E) > 0$ for almost all $E$. In the i.i.d. case under discussion, this follows from Furstenberg's theorem ([**120**]) for *all $E$, but under much weaker conditions ($v$-nondeterministic), one gets the result for a.e. $E$ from Kotani's theory* [**176, 305**].

*3. The theorem of Ruelle-Oseledec* [**263, 230**] *implies that when* (12.4) *holds, with $\gamma(E) > 0$, then there exist initial values $u_\pm(0), u_\pm(1)$ so that the corresponding solutions $u_\pm$ of* (12.3) *are $\ell^2$ at $\pm\infty$. (In fact, the solutions decay exponentially.)*

*4. Theorem 12.4 then implies that if $\gamma(E) > 0$ and* (12.4) *holds, then either $E$ is an eigenvalue of $h_\omega$ or $G(E,\omega) < \infty$ (with $G(E,\omega)$ associated to $h_\omega$ and $\delta_0$).*

*5. By Fubini's theorem, for a.e. $\{v_\omega(n)\}_{n\neq 0}$,* (12.4) *holds for a.e. $E$ with $\gamma > 0$. Since $h_\omega$ has only countably many eigenvalues, we conclude that $G(E,\omega) < \infty$ for a.e. $E$ (for this typical set of $\omega$). Thus, by Theorem 12.5, $h_\omega + \alpha\delta_0$ has pure point spectrum for a.e. $\alpha$.*

*6. By the hypothesis on the absolute continuity of the distribution of $v_\omega(0)$, $h_\omega$ has point spectrum for a.e. $\omega$.* □

Essentially, the identical argument proves

THEOREM 12.7. *Let $\{V_\omega(x)\}_{-\infty<x<\infty}$ be a nondeterministic bounded ergodic process on $\mathbb{R}$. Then for a.e. $\omega$ and a.e. $\theta \in [0,\pi)$,*

$$-\frac{d^2}{dx^2} + V_\omega(x)$$

*on $L^2((0,\infty); dx)$ with $\theta$ the boundary condition at 0 has only pure point spectrum.*

In Chapter 13, we'll need the following formula of Simon-Wolff [**314**]:

THEOREM 12.8. *Suppose $\|\varphi\| = 1$. Let $\delta_n$ be an orthonormal basis for $\mathcal{H}$ with $\delta_0 = \varphi$. Then for each $E \in \mathbb{R}$, $\sum_n |(\delta_n, (A-E-i\epsilon)^{-1}\delta_0)|^2$ is monotone increasing as $\epsilon \downarrow 0$ and*

$$G(E) = \lim_{\epsilon\downarrow 0} \sum_n |(\delta_n, (A-E-i\epsilon)^{-1}\delta_0)|^2$$

PROOF. By the Plancherel theorem, the sum is $\|(A-E-i\epsilon)^{-1}\varphi\|^2 = \int d\mu_0(x)/((x-E)^2+\epsilon^2)$. The monotonicity and identification of the limit are immediate. □

## 12.4. Instability of Point Spectrum

Random potentials have point spectrum embedded in the essential spectrum of $h_\omega$. One of our results in this section is that there are locally uncountably many values of $v_\omega(0)$ which turn that point spectrum into singular continuous spectrum. These are ideas obtained by Gordon [**137, 138**], and then independently by del Rio et al. [**91, 92**], whose presentation we follow.

The key in many ways is Theorem 11.2(iv) which says that $\{x \mid G(x) = \infty\}$ is a dense $G_\delta$. As an example of what this implies, we prove the following (first proven in a slightly weaker form by del Rio [**89**]):

THEOREM 12.9. *Let $V$ be a bounded potential on $[0,\infty)$. Let $S$ be the set of nonisolated points of $\sigma_{\text{ess}}(-\frac{d^2}{dx^2}+V)$ (this set is independent of boundary conditions at 0). Then, there exists $T \subset S$, a dense $G_\delta$ in $S$, so that*

(i) *$E \in T$ is not an eigenvalue of $-\frac{d^2}{dx^2}+V$ for any boundary condition $\theta$ at $x = 0$.*
(ii) *If $E \in T$, $(-\frac{d^2}{dx^2}+V)u = Eu$ does not have a solution $\ell^2$ at infinity and, in particular, there is no Lyapunov behavior with $\gamma(E) > 0$ for $E \in T$.*

PROOF. By Theorem 11.2(iv), the $G(E)$ associated to the Neumann operator with $\varphi = \delta(x)$ has $G(E) = \infty$ on $T_0$, a dense $G_\delta$ in $S$. Let $T$ be $T_0$ minus any eigenvalues for $\theta = \pi/2$. (i), (ii) follow by the results earlier in this chapter. □

EXAMPLE. If $V$ is random with $\text{spec}(H) = [0,\infty)$, this implies non-Lyapunov behavior for a dense $G_\delta$ on $[0,\infty)$, even though we know there is such behavior for a.e. $E$. Dense $G_\delta$ sets are sometimes called Baire generic while complements of sets of measure zero are called Lebesgue generic. Both kinds of sets are locally uncountable. The theorem says that the sets where one does/does not have Lyapunov behavior with $\gamma > 0$ are intertwined, locally uncountable sets!

We want to turn the $G(E) = \infty$ result into one about $\alpha$ values.

THEOREM 12.10 (del Rio et al. [**92**]). *Let $d\mu$ be any measure on $\mathbb{R}$ obeying (11.4). Then the set of values $\{F(E+i0) \mid E \in \text{supp}(d\mu)$ and $G(E) < \infty\}$ is the complement of a dense $G_\delta$.*

PROOF. There is a general argument that the complement is a $G_\delta$ [**308**], so we focus on the fact that the set, $S$, is a countable union of nowhere dense sets. We sketch the idea behind the proof in [**92**], leaving the details for that paper.

Write $\{E \mid\in \text{supp}(d\mu)$ and $G(E) < \infty\}$ as a countable union of sets $A_n$ so that

(i) $G(E)$ is close to the constant, $\gamma_n$, on $A_n$.
(ii) For some $\delta_n$ and all $x \in A_n$, $\int_{|x-y|<\delta_n} |x-y|^{-2}\, d\mu(y)$ is small compared to the variation of $G$ over $A_n$.
(iii) For each $x \in A_n$, consider the equilateral triangle, $T_x$, in $\mathbb{C}$ with one vertex at $x$ and base on the line $\operatorname{Im} z = a\delta_n$ with a fixed number, $a$, (fixed in the proof in [**92**]). Then $\cup_{x\in A_n} T_x$ is connected.

By (i), (ii) $F'(E)$ is very close to the constant $\gamma_n$ on each $T_x$ so $|F(z) - F(y) - \gamma_n(z-y)| \le \frac{1}{3}\gamma_n|y-z|$ on $T_x$. By a simple geometric argument using (iii), $|F(z) - F(y) - \gamma_n(z-y)| \le \frac{2}{3}\gamma_n|y-z|$ for $y, z \in \cup_{x\in A_n} T_x$ and so in $A_n$. Since $A_n$ is nowhere dense and this inequality says that $F$ on $A_n$ is the restriction of a homeomorphism from $\mathbb{R}$ to $\mathbb{R}$, we see that $F[A_n]$ is nowhere dense. □

Combining this with Theorem 12.2, we immediately get

THEOREM 12.11. *In the general context of rank one perturbations, suppose that* $[a,b] \subset \operatorname{spec}(A)$ *and* $\operatorname{spec}_{\rm ac}(A) \cap (a,b) = \emptyset$. *Then for a dense* $G_\delta$ *of* $\alpha$*'s,* $A_\alpha$ *has purely singular continuous spectrum on* $(a,b)$.

We emphasize that there is a distinction between singular and point spectrum. If $\operatorname{spec}_{\rm ac}(A) \cap (a,b) = \emptyset$ and $(a,b) \subset \operatorname{spec}(A)$, there are *always* many $\alpha$'s for which $A_\alpha$ has singular continuous spectrum. There may or may not be $\alpha$'s when there is point spectrum there. The Simon-Wolff criterion gives cases where there is point spectrum for many $\alpha$'s, but the criterion may or may not hold.

REMARK. This result on singular continuous spectrum for dense $G_\delta$'s is one of many recently discovered [**138, 308, 91, 162**]. For example, in [**308**], it is proven that in $C_\infty(\mathbb{R}^n)$, the continuous functions vanishing at infinity, there is a dense $G_\delta$, $S$, so that if $V \in S$, then $-\Delta + V$ has purely singular continuous spectrum on $[0,\infty)$.

COROLLARY 12.12. *Let* $V_\omega(x)$ *be a nondeterministic ergodic random process for* $x \in \mathbb{R}$ *so that* $\operatorname{spec}(-\frac{d^2}{dx^2} + V) = [a,\infty)$ *for some* $a$. *Let* $H_\theta$ *be* $-\frac{d^2}{dx^2} + V$ *on* $L^2(0,\infty)$ *with* $\theta$ *boundary condition at* $x=0$. *Then for a dense* $G_\delta$ *of* $\theta$*'s,* $H_\theta$ *has purely singular continuous spectrum on* $[0,\infty)$ *(while for Lebesgue almost all, it has only point spectrum there).*

COROLLARY 12.13. *Let* $\{V_\omega(n)\}_{n\in\mathbb{Z}^\nu}$ *be independent, identically distributed random variables with uniform distribution in* $[a,b]$ *and let* $H_\omega$ *be the corresponding Jacobi matrix on* $\ell^2(\mathbb{Z}^\nu)$. *Let* $|b-a|$ *be so large that* $H_\omega$ *has only point spectrum for a.e.* $\omega$ *(see Chapter 13). Then for a.e.* $\omega$*, there is a dense* $G_\delta$*,* $S$ *in* $[a,b]$*, so that for* $\alpha \in S$,

$$W(n) = \begin{cases} V_\omega(n) & n \neq 0 \\ \alpha & n = 0 \end{cases}$$

*yields a Jacobi matrix with only singular continuous spectrum.*

Both corollaries follow immediately from the theorem and earlier discussions of the examples.

## 12.5. Examples

The astute reader may have noticed that $d\mu_0$ completely determines the spectral properties of the family $A_\alpha$ (assuming cyclicity, of course). Moreover, given $\mu_0$, let $A$ be multiplication by $x$ on $L^2(\mathbb{R}, d\mu_0)$ and let $\varphi(x) \equiv 1$. Then $F(z)$ is just the Borel transform of $d\mu_0$. So we can describe examples by giving the measure $d\mu_0$, which we'll denote by $d\mu$.

EXAMPLE 1 (The Friedrichs Model). Let $d\mu = dx \restriction [0,1] + \delta_{1/2}$ be the sum of Lebesgue measure on $[0,1]$ and a point mass at $\frac{1}{2}$. Then $G(x) = \infty$ on $[0,1]$, and

$$\frac{1}{\pi} \operatorname{Im} F(x+i0) = \begin{cases} 1 & 0 < x < 1;\ x \neq \frac{1}{2} \\ \frac{1}{2} & x = 0, 1 \end{cases}$$

Thus, $A_\alpha$ has only a.c. spectrum on $[0,1]$ for $\alpha \neq 0$ (there is a single eigenvalue on $(-\infty,0)$ for $\alpha < 0$ and on $(1,\infty)$ for $\alpha > 0$). The embedded eigenvalue dissolves.

EXAMPLE 2. Let $\{q_n\}_{n=1}^{\infty}$ be a counting of the rationals on $[0,1]$. Let

$$d\mu = dx \restriction [0,1] + \sum_{n=1}^{\infty} 2^{-n}\delta_{q_n}$$

As above, all eigenvalues disappear for $\alpha \neq 0$. The interesting case is to take $d\mu = d\mu_{-1}$. The new $A$ has no eigenvalues on $[0,1]$ but $A + (\varphi, \cdot\,)\varphi$ suddenly has dense point spectrum embedded in the a.c. spectrum.

EXAMPLE 3. Let $d\mu$ be conventional Cantor measure. We claim that $G(x) = \infty$ on $C$, the Cantor set. In fact,

$$\operatorname{Im} F(x + i\epsilon) \to \infty \tag{12.5}$$

for $x \in C$, which implies that $G(x) = \infty$. This fact about $\operatorname{Im} F$ implies that $A_\alpha$ has neither point nor singular spectrum on $C$. In fact, each $A_\alpha$ has one eigenvalue $e_j(\alpha)$ in each interval, $I_j$ in $[0,1]\backslash C$. The corresponding eigenvectors, together with one, $e_0(\alpha)$, in $I_0 \equiv \mathbb{R}\backslash[0,1]$, are complete. $\operatorname{spec}(A_\alpha) = \cup_{j=0}^{\infty}\{e_j(\alpha)\} \cup C$ since the limit points of the $e$'s are precisely $C$. To prove (12.5), we claim first that

$$\operatorname{Im} F(x_0 + i\epsilon) \geq \frac{1}{2\epsilon}\,\mu(\{y \mid |x - y| \leq \epsilon\}) \tag{12.6}$$

for any measure $\mu$. Moreover, if $x \in C$, then

$$\mu(\{y \mid |x - y| \leq 2(3^{-n})\}) \geq 2^{-n}$$

so (12.6) implies (12.5).

EXAMPLE 4. (Simon-Wolff [**314**]) Let $d\mu_n = \frac{1}{2^n}\sum_{j=1}^{2^n}\delta_{j/2^n}$ and $\mu = \sum a_n\, d\mu_n$ where $\sum a_n < \infty$. If $x \in [0,1]$, then clearly, since there is a $j$ with $|x - (j/2^n)| \leq 2^{-n}$, we have

$$\int |x - y|^{-2}\, d\mu_n(y) \geq (2^{-n})^{-2}2^{-n} = 2^n$$

so if $\sum 2^n a_n = \infty$, then $G(x) = \infty$ on $[0,1]$ and thus $A_\alpha$ has no point spectrum $[0,1]$. So $A$ has dense point spectrum on $[0,1]$ which turns into purely singular continuous spectrum on $[0,1]$ for all $\alpha \neq 0$.

EXAMPLE 5. Let $d\nu$ be Cantor measures and let

$$d\mu(x) = \chi_{[0,1]}\sum_{n=1}^{\infty} n^{-2}\left[\sum_{j=1}^{2^n} d\nu(x - (j/2^n))\right]$$

Then $\mu\{y \mid |y - x| \leq 2^{-n+1}\} \geq 3^{-n}n^{-2}$ for any $\chi \in [0,1]$ so $G(x) \geq \frac{1}{4}(2^{-n})^{-2}3^{-n}n^{-2} = \frac{1}{4}(\frac{4}{3})^n n^{-2}$ for any $n$ and so $G = \infty$ on $[0,1]$. This yields an example with $\sigma(A_\alpha)$ purely singular continuous on $[0,1]$ for all $\alpha$. As noted, this cannot happen for point spectrum.

EXAMPLE 6. (Simon-Wolff [**314**]) Let $x_n$ be arbitrary points in $[0,1]$. Let $a_n$ be a sequence with $0 \leq a_n \leq C_0\alpha^n$ for some $\alpha < 1$. Let

$$d\mu = \sum a_n\delta_{x_n}$$

We claim $G(x) < \infty$ for a.e. $x$. For if $|x - x_n| > D^{1/2}\alpha^{n/4}$ for *all* $n$, then

$$G(x) < \sum_n C_0\alpha^n D^{-1}\alpha^{-n/2} = \frac{C_0}{D(1 - \alpha^{1/2})}$$

Thus

$$\left|\left\{x \,\middle|\, G(x) > \frac{C_0}{D(1-\alpha^{1/2})}\right\}\right| \leq \sum 2D^{1/2}\alpha^{n/4} = 2D^{1/2}(1-\alpha^{1/2})^{-1}$$

goes to zero as $D \to 0$, proving that $G(x) < \infty$ for a.e. $x$. In this case, by the Simon-Wolff criterion, $A_\alpha$ has only pure point spectrum for a.e. $\alpha$, but for a dense $G_\delta$ of $\alpha$, it has singular continuous spectrum, of course.

Let $x_n$ be the counting of the dyadic rationals. Let $d\mu = \sum b_n \delta_{x_n}$. Examples 4 and 6 say that depending on the $b$'s, $A_\alpha$ can have dense point or singular continuous spectrum for Lebesgue typical $\alpha$. Spectral properties depend heavily on $\varphi$ as well as $A$!

EXAMPLE 7. (del Rio et al. [**93**]) Let $\{q_n\}_{n=1}^{\infty}$ be a counting of the rationals in $(0,1)$. Define $a_n = \min(3^{-n-1}, q_n, 1-q_n)$ and let $S = \cup_n(q_n - a_n, q_n + a_n)$ so $|S| \leq 2\sum a_n \leq \frac{1}{3}$. Let $d\mu_0 = \chi_S\, dx$ so $\operatorname{spec}(A_0) = [0,1]$ since $S$ is dense in $[0,1]$. Since $|[0,1]\backslash S| \geq \frac{2}{3}$ and $d\mu_{\alpha,\mathrm{ac}}([0,1]\backslash S) = 0$, there must be a positive measure set of $\alpha$'s with point and/or singular spectrum embedded in $\operatorname{spec}(A_\alpha)$. As in the last example, if $|x - q_n| \geq a_n + \sqrt{\epsilon}\,3^{-n/4}$ for all $n$, then $G(x) \leq \epsilon^{-1}(1-\sqrt{1/3})^{-1}$. Thus, $\{x \in S \mid G(x) > \epsilon^{-1}(1-\sqrt{1/3})^{-1}\} = \sqrt{\epsilon}\,(1-\sqrt{1/3})^{-1}$ so $G(x) < \infty$ for a.e. $x \in [0,1]\backslash S$. We conclude that for a.e. $\alpha$, $A_\alpha$ has no singular spectrum on $S$, but it has some point spectrum there for a positive measure set of $\alpha$'s.

EXAMPLE 8. (del Rio et al. [**93**]) Let $S_n = \cup_{j=1}^{n-1}(\frac{j}{2^n} - \frac{1}{4n^2 2^n}, \frac{j}{2^n} + \frac{1}{4n^2 2^n})$ and $S = \cup_n S_n$, so $|S_n| \leq \frac{1}{2n^2}$ and $|S| \leq \frac{\pi^2}{12} < 1$. Let $d\mu_0 = \chi_S\, dx$. Then, $\operatorname{spec}(A_0) = [0,1]$ but since $|[0,1]\backslash S| > 0$, there must be a positive measure set of $\alpha$'s with point and/or singular spectrum embedded in $\operatorname{spec}(A_\alpha)$. As in Example 4, $G(x) = \infty$ on $[0,1]$ since for any $n$,

$$G(x) \geq (2^{-n-1})^{-2} 2^{-n-1} n^{-2} = \frac{2^n}{8n^2} \to \infty$$

Thus, for all $\alpha$, we have no point spectrum in $[0,1]$ but we have singular continuous spectrum embedded in the a.c. spectrum for a set of positive measure $\alpha$'s.

CHAPTER 13

# Localization in the Anderson Model Following Aizenman-Molchanov

In this chapter, we'll present the approach of Aizenman-Molchanov [**8**] to the proof of localization for the large coupling Anderson model in arbitrary dimension. The proof is so simple, we won't even put it in separate sections. Results on exponential decay of Green's functions were first obtained by Fröhlich-Spencer [**119**], with later extensions and simplifications by [**117, 336**]. Even taking into account the fact that we have already developed some of the facts we need, this proof is something like a factor of ten shorter and has smaller constants.

We will prove:

THEOREM 13.1. *Let* $\{V_\omega(n)\}_{n\in\mathbb{Z}^\nu}$ *be independent, identically distributed random variables with uniform density on* $[0,1]$. *Then, there is a fixed constant,* $K$, *so that if* $|\lambda| > K\nu^2$, *then the Jacobi matrix,* $h_0 + \lambda V_\omega$, *has only pure point spectrum (and it is dense in* $[-2\nu, 2\nu+\lambda]$).

REMARK. I believe that $K = 32$ is what comes out of the minimization problem below.

Suppose for a moment that we could prove that $E_\omega(\sum_n |G(n,0;E+i0)|^2) \leq Ce^{-D|n|}$. Then, by Fubini's theorem, we would get that for a.e. pair $(E,\omega)$, $\sum_n |(\delta_n,(A-E-i0)^{-1}\delta_0)|^2 < \infty$, so by Theorems 12.8 and 12.5, we would have localization. The problem with this idea is that there is no chance that even $E_\omega(|G(0,0;E+i0)|^2)$ is finite, since after all, $G(0,0;E+i0)$ is precisely $F_\alpha(E+i0)$ for $\alpha = V_\omega(0)$, and that is $F/[1+\alpha F]$ and has a first-order zero at $\alpha = -1/F$ if there is to be an eigenvalue. But then, $\int dv(0)(v(0)-a)^{-2}$ is infinite if $a \in \operatorname{supp}(v(0))$. So the first key idea of [**8**] is to look at powers of $|G|^s$ with $0 < s < 1$. For simplicity, we will take $s = \frac{1}{2}$. The following lemma will be useful:

LEMMA 13.2. *If* $s > 1$, *then for* $\{x_n\}_{n=1}^N$ *all positive* $(N = 1,2,\ldots,\infty)$,

$$\sum_{n=1}^N {x_n}^s \leq \left(\sum_{n=1}^N x_n\right)^s$$

*If* $s < 1$, *then*

$$\sum_{n=1}^N {x_n}^s \geq \left(\sum_{n=1}^N x_n\right)^s$$

PROOF. If $s > 1$, ${x_n}^{s-1} \leq (\sum_{m=1}^N x_m)^{s-1}$,

$$ {x_n}^s = x_n {x_n}^{s-1} \leq x_n \left(\sum_{m=1}^N x_m\right)^{s-1}$$

Now sum over $n$. If $s < 1$, ${x_n}^{s-1} \geq (\sum_{m=1}^N x_m)^{s-1}$ and the proof is identical. □

In particular,

$$\left(\sum_n |G(n,0;E+i\epsilon)|^2\right)^{1/4} \leq \sum_n |G(n,0;E+i\epsilon)|^{1/2} \tag{13.1}$$

So we need only control $\sum_n E_\omega(|G(n,0;E+i\epsilon)|^{1/2})$.

A key lemma that we will need is:

LEMMA 13.3. *There is a universal constant, $C_0$, so that for all complex numbers, $\alpha, \beta$,*

$$\int_0^1 |x-\alpha|^{1/2}|x-\beta|^{-1/2}\,dx \geq C_0 \int_0^1 |x-\beta|^{-1/2}\,dx \tag{13.2}$$

REMARK. I believe $\alpha = \beta = \frac{1}{2}$ is the extreme point, in which case $C_0 = 1/\sqrt{8}$.

PROOF. It is easy to see the integral on the left side of (13.2) only decreases if $\alpha$ is replaced by $\operatorname{Re}\alpha$ or if $\alpha \notin [0,1]$ and is replaced by the nearest point in $[0,1]$. So we can suppose that $\alpha \in [0,1]$. By $x \to \frac{1}{2} - x$ symmetry, we can suppose $\alpha \in [0,\frac{1}{2}]$. But then

$$\int_0^1 |x-\alpha|^{1/2}|x-\beta|^{-1/2}\,dx \geq \sqrt{1/4}\int_{3/4}^1 |x-\beta|^{-1/2}\,dx$$

so we need only show that

$$\inf_{\beta\in\mathbb{C}}\left[\frac{\int_{3/4}^1 |x-\beta|^{-1/2}\,dx}{\int_0^1 |x-\beta|^{-1/2}\,dx}\right] > 0$$

Let $h(\beta)$ be the function in $[\,\cdot\,]$. It is clearly continuous on $\mathbb{C}$, nonvanishing, and $\lim_{\beta\to\infty} h(\beta) = \frac{1}{4}$, so the inf is strictly positive. □

One last lemma we will need concerns the free Laplacian, $h_0$.

LEMMA 13.4. (a) *Let $f, g \in \ell^\infty(\mathbb{Z}^\nu)$ be nonnegative functions and suppose that*

$$(1 - ah_0)f \leq g$$

*for some $0 < a < 1/2\nu$. Then*

$$f \leq (1-ah_0)^{-1}g$$

(b) *If $0 < a < (2\nu)^{-1}$, then*

$$(1-ah_0)^{-1}(i,j) \leq (2\nu a)^{|j-i|}(1-2\nu a)^{-1}$$

PROOF. $h_0$ has norm $2\nu$ so $(1-ah_0)^{-1} = \sum_{m=0}^\infty (ah_0)^m$ if $2\nu a < 1$. This is a positivity-preserving operator, so it preserves inequalities, which proves (a). Moreover, by the above expression and $(ah_0)^m(i,j) = 0$ if $m < |i-j|$, we see that

$$(1-ah_0)^{-1}(i,j) = \sum_{m=|i-j|}^\infty (ah_0)^m(i,j) \leq \sum_{m=|i-j|}^\infty (2\nu a)^m$$

proving (b). □

PROOF OF THEOREM 13.1. Fix $z \in \mathbb{C}\backslash\mathbb{R}$ and let $g_\omega(n) = G_\omega(0, n; z)$ and

$$k(n) = E_\omega(|g_\omega(n)|^{1/2})$$

We claim that (with $C_0$ the constant in Lemma 13.3)

$$k(n) \le (C_0\lambda^{1/2})^{-1}\bigg[\bigg(\sum_{|j|=1} k(n+j)\bigg) + \delta_{n0}\bigg] \tag{13.3}$$

For $g_\omega$ obeys

$$\sum_{|j|=1} g_\omega(n+j) + (\lambda v_\omega(n) - z)g_\omega(n) = \delta_{n0}$$

Using Lemma 13.2,

$$E_\omega(|\lambda v_\omega(n) - z|^{1/2}|g_\omega(n)|^{1/2}) \le \delta_{n0} + \sum_{|j|=1} k(n+j) \tag{13.4}$$

Fix $\{v_\omega(\ell)\}_{\ell\neq n}$. Then, by (11.14), $g_\omega(n) = C/(\lambda v_\omega(n) + d)$ with $C \neq 0$ and $C, d$ are functions of $\{v_\omega(\ell)\}_{\ell\neq n}$. Thus averaging only over $v_\omega(n)$,

$$\begin{aligned}
\int_0^1 |\lambda v - z|^{1/2}\left|\frac{C}{\lambda v + d}\right|^{1/2} dv &= \lambda^{1/2}\int_0^1 \left|v - \frac{z}{\lambda}\right|^{1/2}\left|\frac{C}{\lambda v + d}\right|^{1/2} dv \\
&\ge \lambda^{1/2}\left(\frac{C}{\lambda}\right)^{1/2}\int_0^1 \left|v - \frac{z}{\lambda}\right|^{1/2}\left|\frac{1}{v + \frac{d}{\lambda}}\right|^{1/2} dv \\
&= C_0\lambda^{1/2}\int_0^1 \left|\frac{C}{\lambda v + d}\right|^{1/2} dv
\end{aligned}$$

where $C_0$ is the constant in Lemma 13.3. Averaging over $\{v_\omega(\ell)\}_{\ell\neq n}$, we get

$$E_\omega(|\lambda v_\omega - z|^{1/2}|g_\omega(n)|^{1/2}) \ge C_0\lambda^{1/2}k(n) \tag{13.5}$$

Obviously, (13.4)/(13.5) imply (13.3).

(13.3) can be rewritten to say

$$(1 - C_0\lambda^{1/2}h_0)k \le \delta_{n0}$$

Because $z \notin \mathbb{R}$, $|g_\omega(n)| \le (1/|\mathrm{Im}\, z|)^{1/2}$, so $k \in \ell^\infty$. We can therefore apply Lemma 13.4 to conclude that if $(C\lambda^{1/2})^{-1} < (2\nu)^{-1}$ (or $\lambda > 4C^{-2}\nu^2$),

$$\begin{aligned}
k(n) &\le (C_0\lambda^{1/2})^{-1}(1 - (C\lambda^{1/2})^{-1}h_0)^{-1}(0, n) \\
&\le (C_0\lambda^{1/2})^{-1}[(2\nu)(C_0\lambda^{1/2})^{-1}]^{|n|}(1 - 2\nu(C_0\lambda^{1/2})^{-1})^{-1}
\end{aligned}$$

Thus, $\sum_n k(n)$ is bounded uniformly in $z$, so by (13.1),

$$\sup_z E_\omega\bigg(\bigg(\sum_n |G(n, 0; z)|^2\bigg)^{1/4}\bigg)$$

is bounded uniformly in $z$. Using Theorem 12.8, this implies that for $A = h_\omega$, $\varphi = \delta_0$, we have

$$E_\omega(G(E)^{1/4}) < \infty$$

for all $E \in \mathbb{R}$. Thus for a.e. pairs, $(V, E)$, $G(E) < \infty$. Now apply Theorem 12.5 to get localization. □

Actually, the proof shows that $E_\omega((e^{\delta|n|}\sum_n |G(n,0;z)|^2)^{1/4}) < \infty$ for some $\delta > 0$, and this implies exponential decay of eigenfunctions (essentially by (11.14)). Indeed, one easily sees that each eigenvector $\varphi$ obeys $|\varphi(n)| \le C_{\varphi,\epsilon}[\frac{\nu^2}{k|\lambda|}]^{(1-\epsilon)|n|/2}$ for each $\epsilon > 0$. For extensions of these ideas, see [**8, 4**]; see also Addendum P.

CHAPTER 14

# The Xi Function

In this chapter, I'll discuss a program of F. Gesztesy and mine (in part with Holden and Zhao) that deals with inverse spectral theory for one-dimensional Schrödinger and Jacobi matrices [**129**]. In the study of the inverse problem for periodic potentials, an important role is played by the trace formula originally proven by Gel'fand-Levitan and exploited/refined by many authors subsequently; see [**132**] for a history.

Let me describe this formula: let $V(x+L) = V(x)$. Then $\operatorname{spec}(-\frac{d^2}{dx^2}+V(x)) = [E_0, E_1] \cup [E_2, E_3], \dots$ where $E_0, E_3, E_4, \dots, E_{4n-1}, E_{4n}, \dots$ are periodic eigenvalues, and $E_1, E_2, \dots, E_{4n+1}, E_{4n+2}, \dots$ are antiperiodic eigenvalues. Let $\{\mu_j(x)\}_{j=1}^\infty$ be the eigenvalues on $L^2(x, x+L)$ with $u(x) = u(x+L) = 0$ Dirichlet boundary conditions. It can be shown that $E_{2j-1} \le \mu_j(x) \le E_{2j}$. The trace formula says that, at least when $V$ is smooth enough ($C^1$ will do),

$$V(x) = E_0 + \sum_{j=1}^\infty (E_{2j} + E_{2j-1} - 2\mu_j(x)) \tag{14.1}$$

One of our goals in this chapter is to extend this formula to more general $V$'s than periodic. One issue to consider immediately is a case like $V(x) = x^2 - 1$ where $V(x) \to \infty$ as $|x| \to \infty$. Then the bands, $[E_{2j}, E_{2j+1}]$ collapse to a point and we have eigenvalues $E_0 < E_1 < \cdots$ and Dirichlet eigenvalues $\{\mu_j(x)\}_{j=1}^\infty$ (with $u(x) = 0$ boundary conditions) and $E_{j-1} \le \mu_j(x) \le E_j(x)$. The formal analog of (14.1) is

$$V(x) = E_0 + \sum_{j=1}^\infty (E_{j-1} + E_j - 2\mu_j(x)) \tag{14.2}$$

As an example, take $V(x) = x^2 - 1$ and look at $\mu_j(x=0)$: $E_n = 2n$, $n = 0, 1, 2, \dots$, and $\{\mu_j(0)\}_{j=1}^\infty$ are $\{2, 2, 6, 6, 10, 10, \dots\}$. It follows that (14.2) says

$$-1 = -2 + 2 - 2 \dots$$

Since the Abelian sum $1 - 1 + 1 - \cdots$ is $\frac{1}{2}$, this is promising — but it is clear that the final result will need a summability method to accommodate the formula. Indeed, our trace formula (when $V > 0$) will read: Let $\xi(x, \lambda)$ be the Krein spectral shift for infinite coupling with $A = -\frac{d^2}{dx^2} + V(x)$ and $\varphi = \delta(x)$. Then

$$V(x) = \lim_{\gamma\to\infty} \int_0^\infty \frac{\gamma^2}{(\lambda+\gamma)^2} (1 - 2\xi(x, \lambda))\, dx \tag{14.3}$$

Note that $|1 - 2\xi(x, \lambda)| = 1$ in all discrete spectrum cases, clearly illustrating the need for a summability method.

## 14.1. Abstract Trace Formula

We recall some facts from Sections 11.5 and 11.6. As usual, let $A \geq 0$, $A_\alpha = A + \alpha(\varphi, \cdot\,)\varphi$ with $A_\infty$ defined as the strong resolvent limit of $A_\alpha$ as $\alpha \to \infty$. Define

$$F(z) = (\varphi, (A - z)^{-1}\varphi) \tag{14.4}$$

and $\xi_\infty$ by

$$\xi_\infty(\lambda) = \frac{1}{\pi} \arg F(\lambda + i0) \tag{14.5}$$

Then

$$\operatorname{Tr}(f(A) - f(A_\infty)) = -\int f'(\lambda)\xi_\infty(\lambda)\, dx \tag{14.6}$$

and

$$\int_0^\infty \frac{\xi_\infty(\lambda)\, d\lambda}{(\lambda - z)^2}\, dx = \frac{d}{dz} \log F(z) \tag{14.7}$$

We will also consider a second problem with $A^{(0)}$, $\xi_\infty^{(0)}$, and $F^{(0)}$. The point of a comparison is that, with great generality, Green's functions, $G(0, 0; z)$, have the same behavior, namely, $(2(-z)^{1/2})^{-1}$ at $z = -\infty$.

THEOREM 14.1 (Gesztesy-Simon [**132**]). *Suppose that* $\lim_{\gamma\to\infty} \frac{F(-\gamma)}{F^{(0)}(-\gamma)} = 1$. *Then for* $z \in \mathbb{C}\backslash[0, \infty)$,

$$F(z) = F^{(0)}(z) \exp\biggl(\lim_{\gamma\to\infty} \int \frac{\gamma}{\lambda + \gamma}\, \frac{1}{\lambda - z}\, (\xi_\infty - \xi_\infty^{(0)})(x)\, dx\biggr) \tag{14.8}$$

PROOF. (14.7) for $F, F^{(0)}$ integrated says that

$$\log\biggl(\frac{F(z)}{F^{(0)}(z)}\biggr) - \log\biggl(\frac{F(-\gamma)}{F^{(0)}(-\gamma)}\biggr) = \int \biggl(\frac{1}{\lambda - z} - \frac{1}{\lambda + \gamma}\biggr)(\xi_\infty - \xi_\infty^{(0)})(\lambda)\, d\lambda \tag{14.9}$$

Moreover, by dominated convergence, for $z$ fixed,

$$\lim_{\gamma\to\infty} \int \frac{|z|}{|\lambda - z|}\, \frac{1}{\lambda + \gamma}\, |(\xi_\infty - \xi_\infty^{(0)})(\lambda)|\, d\lambda = 0 \tag{14.10}$$

(14.9)–(4.10) and $\lim_{\gamma\to\infty} \log(\frac{F(-\gamma)}{F^{(0)}(-\gamma)}) = 0$ imply (14.8). □

THEOREM 14.2 (The Abstract Trace Formula [**132**]). *Suppose that for some* $\varepsilon > 0$,

$$\lim_{\substack{|\gamma|\to\infty \\ |\arg\gamma|<\varepsilon}} \gamma\biggl(\frac{F(-\gamma)}{F^{(0)}(-\gamma)} - 1\biggr) = a$$

*Then*

$$a = \lim_{\substack{\gamma\to\infty \\ \gamma \text{ real}}} \int \frac{\gamma^2}{(\lambda + \gamma)^2} (\xi_\infty - \xi_\infty^{(0)})(\lambda)\, d\lambda$$

PROOF. By hypothesis,

$$\lim_{\substack{|\gamma|\to\infty \\ |\arg\gamma|<\varepsilon}} \gamma \log\biggl[\frac{F(-\gamma)}{F^{(0)}(-\gamma)}\biggr] = a$$

By a simple contour argument,

$$\lim_{\substack{\gamma\to\infty \\ \gamma \text{ real}}} \gamma^2 \biggl(\frac{d}{dz} \log\biggl[\frac{F(z)}{F^{(0)}(z)}\biggr]\biggr)_{z=-\gamma} = a$$

This plus (14.8) implies the result. □

## 14.2. The Trace Formula for Schrödinger Operators

We will postpone the proof of the following until later in this section:

LEMMA 14.3. *Suppose that $V$ is bounded from below and is continuous at $x$. Let $G$ be the Green's function for $-\frac{d^2}{dx^2}+V(x)$ on $L^2(-\infty,\infty)$. Then as $|\gamma|\to\infty$ with $|\arg\gamma|<\pi/2$:*

$$G(x,x;-\gamma)=\tfrac{1}{2}\,(\gamma)^{-1/2}(1-\tfrac{1}{2}\,\gamma^{-1}V(x)+o(\gamma^{-1})) \tag{14.11}$$

EXAMPLE. This will be the comparison $F^{(0)}$ so it is a critical example: Let $A^{(0)}=-\frac{d^2}{dx^2}$ on $L^2(\mathbb{R},dx)$. Let $\varphi=\delta_x$. Then $F^{(0)}(z)=G^{(0)}(x,x;z)=\frac{1}{2}(-z)^{-1/2}$. Thus $F^{(0)}(x+i0)=\frac{1}{2}(x)^{-1/2}i$ and $\xi_\infty^{(0)}(x)=\frac{1}{2}$, $x>0$.

We now apply the abstract theory of the last section:

THEOREM 14.4. *Suppose that $V$ is nonnegative and is continuous at $x$. Let $\xi(x,\lambda)$ be $\xi_\infty(x)$ for $A=-\frac{d^2}{dx^2}+V$; $\varphi=\delta_x$. Then*

$$V(x)=\lim_{\gamma\to\infty}\int_0^\infty\frac{\gamma^2}{(\lambda+\gamma)^2}\,[1-2\xi(x,\lambda)]\,d\lambda \tag{14.12}$$

*and*

$$G(x,x;z)=\frac{1}{2}\,(-z)^{-1/2}\exp\biggl(\lim_{\gamma\to\infty}\int_0^\infty\frac{\gamma}{\gamma+\lambda}\,\frac{1}{\lambda-z}\biggl[\xi(x,\lambda)-\frac{1}{2}\biggr]dx\biggr) \tag{14.13}$$

PROOF OF LEMMA 14.3 WHEN $V$ IS BOUNDED. Without loss of generality, suppose $x=0$. Let $\psi(x)=(H_0+\gamma)^{-1}\delta_0(x)=\frac{1}{2}\gamma^{-1/2}e^{-\gamma^{1/2}|x|}$, so $\|\psi_\gamma\|=0(\gamma^{-3/4})$. By the second resolvent equation, since $V$ is bounded,

$$(\delta_0,(H_0+V+\gamma)^{-1}\delta_0)=(\delta_0,(H_0+\gamma)^{-1}\delta_0)-(\psi_\gamma,V\psi_\gamma)+R \tag{14.14}$$

where $\|R\|\le\|\psi_\gamma\|^2\gamma^{-1}=0(\gamma^{-5/2})$. The first term in (14.14) is $\frac{1}{2}\gamma^{-1/2}$ and the second is

$$-\tfrac{1}{4}\,\gamma^{-1}\int_{-\infty}^\infty e^{-2\gamma^{1/2}|x|}V(x)\,dx=-\tfrac{1}{4}\,\gamma^{-3/2}[V(0)+o(1)]$$

since $\int_{-\infty}^\infty e^{-2\alpha|x|}\,dx=\alpha^{-1},\alpha>0$. □

PROOF OF LEMMA 14.3 IN THE GENERAL CASE. We will prove with $H=H_0+V$ that

$$e^{-tH}(x,x)=(4\pi t)^{-1/2}(1-tV(x)+o(t)) \tag{14.15}$$

(14.11) then follows from

$$G(x,x;-\gamma)=\int_0^\infty e^{-\gamma s}e^{-sH}(x,x)\,ds$$

and

$$\int_0^\infty t^{-1/2}e^{-t}\,dt=\sqrt{\pi}\qquad\text{and}\qquad\int_0^\infty t^{1/2}e^{-t}\,dt=\tfrac{1}{2}\sqrt{\pi}$$

To prove (14.15), let $\{\omega(s)\}_{0\le s\le 1}$ be the Brownian bridge [**303**], so the Feynman-Kac formula has the form [**303**]

$$e^{-tH}(x,x)=(4\pi t)^{-1/2}E_\omega\biggl(\exp\biggl[-t\int_0^1 V(x+2\sqrt{2t}\,\omega(s))\,ds\biggr]\biggr) \tag{14.16}$$

A Levy-type estimate implies that

$$E_\omega\big(\sup_{0\le s\le 1}\omega(s)\ge\alpha\big)\le C_1e^{-C_2\alpha^2}$$

Changing $V$ outside $(x-\delta, x+\delta)$ changes the right side of (14.16) by $0(e^{-C/t})$, so in proving (14.15), we can suppose that $V$ is bounded. But if $V$ is bounded, it is easy to see that the integral is $1-tV(x)+o(t)$. □

Since we controlled $e^{-tH}(x,x)$, we can prove the slightly stronger version of (14.12), namely [**132**],

$$V(x)=\lim_{\gamma\downarrow\infty}\int_0^\infty e^{-\gamma\lambda}(1-2\xi(x,\lambda))\,d\lambda$$

This is all with $V$ positive. If it is only bounded below and $E_0\le\inf\operatorname{spec}(H)$, then

$$V(x)=E_0+\lim_{\gamma\downarrow 0}\int_{E_0}^\infty e^{-\gamma\lambda}(1-2\xi(x,\lambda))\,d\lambda$$

See [**130**] for higher-order asymptotic formulae involving moments of $\xi$.

## 14.3. Examples

EXAMPLE 1 (Periodic Potentials). We begin with the motivating case of periodic potentials, $V(x)=V(x+L)$. The spectrum of $H$ is $[E_0,E_1]\cup[E_2,E_3]\cup\cdots$. General principles imply that $\xi(x,\lambda)$ obeys

$$\xi(x,\lambda)=\begin{cases}\frac12 & E_{2n}<\lambda<E_{2n+1}\\ 1 & E_{2n-1}<\lambda<\mu_n(x)\\ 0 & \mu_n(x)<\lambda<E_{2n}\text{ or }\lambda<E_0\end{cases}$$

If $V$ is smooth enough (e.g., $V\in C^1$), $\sum|E_{2n}-E_{2n-1}|<\infty$, so $\int|2-2\xi(x,\lambda)|\,d\lambda<\infty$ and the summability method is not needed. Clearly,

$$\int_{E_{2n-1}}^{E_{2n}}(1-2\xi)\,d\lambda=-(\mu_n-E_{2n-1})+(E_{2n}-\mu_n)=E_{2n}+E_{2n-1}-2\mu_n$$

$$\int_0^{E_0}(1-2\xi)\,d\lambda=E_0$$

and we get

$$V(x)=E_0+\sum_{n=1}^\infty(E_{2n}+E_{2n-1}-2\mu_n(x))$$

EXAMPLE 2 (Harmonic Oscillator). $V(x)=x^2-1$; the eigenvalues are $0,2,4\ldots$ and

$$\xi(0,\lambda)=\begin{cases}1 & 0<\lambda<2;\ 4<\lambda<6,\ldots\\ 0 & 2<\lambda<4;\ 6<\lambda<8,\ldots\end{cases}$$

$$\begin{aligned}\int_0^\infty e^{-\alpha\lambda}(1-2\xi(0,\lambda))\,d\lambda&=\alpha^{-1}(1-e^{-2\alpha})\sum_{n=0}^\infty(-1)^{n+1}e^{-2n\alpha}\\&=\alpha^{-1}(1-e^{-2\alpha})(-1)/[1+e^{-2\alpha}]\end{aligned}$$

converges to $-1$ as $\alpha\downarrow 0$. Thus the summability method yields $V(0)$. In general,

THEOREM 14.5. *Let $V(x) \to \infty$ as $|x| \to \infty$. Let $E_0 \le \mu_1(x) \le E_1 \le \cdots$ be the eigenvalues of $-\frac{d^2}{dx^2} + V$ on all of $L^2(\mathbb{R}, dx)$ and with a Dirichlet boundary condition at $x$. Then*

$$V(x) = E_0 + \lim_{\gamma \downarrow 0} \sum_{n=1}^{\infty} [2e^{-\gamma\mu_n(x)} - e^{-\gamma E_n} - e^{-\gamma E_{n-1}}]\gamma^{-1}$$

EXAMPLE 3 (Short-Range Potentials). Suppose $|V(x)| \le C(1+|x|)^{-1-\epsilon}$. Let $R_\ell(\lambda)$ be the usual reflection coefficient and $f_\ell(x, \lambda)$ the Jost function at $-\infty$. Then [**128**]

$$\xi(x, \lambda) = \frac{1}{2} + \arg\left[\frac{1 + R_\ell(\lambda) f_\ell(x, \lambda)^2}{|f_\ell(x, \lambda)|^2}\right] \qquad \lambda > 0$$

We see that if $\int |R_\ell(\lambda)|\, d\lambda < \infty$, then $\int |1 - 2\xi(x, \lambda)|\, dx < \infty$, and one can take the limit inside the integral. See [**128**] for other short-range examples.

As a preliminary to the next example, we need a formula about how Dirichlet eigenvalues $\mu(x)$ move as $x$ changes.

THEOREM 14.6. *Let $(a, b)$ be an interval in $\mathbb{R}\backslash\mathrm{spec}(H)$ (i.e., a spectral gap of $H$) and suppose $\mu(x)$ is an eigenvalue of $H_x$, the operator with a Dirichlet boundary condition at $x$ with $\mu(x) \in (a, b)$. Let $g(x, \lambda) \equiv G(x, x; \lambda)$. Then*

(i) $\frac{\partial g}{\partial x}(x, \lambda)\Big|_{\lambda = \mu(x)} = \pm 1$ *with the value equal $+1$ (resp. $-1$) if the eigenvector lives on $(-\infty, 0]$ (resp. $[0, \infty)$).*

(ii) *With the same sign as in* (i),

$$\mu'(x) = \pm\left(\frac{\partial g}{\partial \mu}(x, \mu(x))\right)^{-1} \tag{14.17}$$

PROOF. (i) By general principles with $\varphi_\pm$ $L^2$ at $\pm\infty$,

$$g(x, \lambda) \equiv \frac{\varphi_-(x)\varphi_+(x)}{\varphi_+(x)\varphi_-'(x) - \varphi_-(x)\varphi_+'(x)}$$

so

$$\frac{\partial g}{\partial x}(x, \lambda) = \frac{\varphi_+(x)\varphi_-'(x) + \varphi_-(x)\varphi_+'(x)}{\varphi_+(x)\varphi_-'(x) - \varphi_-(x)\varphi_+'(x)}$$

At $\mu(x)$, either $\varphi_+(x) = 0$, in which case $\frac{\partial g}{\partial x} = -1$, or $\varphi_-(x) = 0$, in which case $\frac{\partial g}{\partial x} = 1$ as claimed.

(ii) Immediate from (i), given the implicit function theorem that says that $\mu' = -[\frac{\partial g}{\partial x}]/[\frac{\partial g}{\partial \mu}]$. □

EXAMPLE 4 (The Soliton). A *reflectionless* potential is one with $\xi(x, \lambda) = \frac{1}{2}$ for *all* $x$ and a.e. $\lambda$ in $\mathrm{spec}_{\mathrm{ac}}(H)$. Let us look for reflectionless potentials with a single eigenvalue at $E = -1$. Let $\mu(x)$ be the Dirichlet eigenvalue in $(-1, 0]$. Then

$$\xi(x, \lambda) = \begin{cases} 0 & \lambda < -1 \text{ or } \mu(x) < \lambda < 0 \\ 1 & -1 < \lambda < \mu(x) \\ \frac{1}{2} & \lambda > 0 \end{cases}$$

The formulae for $g$ and $V$ then equal

$$V(x) = 2(-1 - \mu(x))$$

$$g(x,\lambda) = \tfrac{1}{2}\,(-\lambda)^{-1/2}\,\frac{\mu(x)-\lambda}{-1-\lambda}$$

(14.17) reads

$$\mu'(x) = \pm 2(-\mu)^{1/2}(1+\mu) \tag{14.18}$$

Let $c(x) = (-\mu)^{1/2}$. Then (14.18) becomes

$$c' = \pm(1-c^2)$$

whose solution is $c(x) = \pm\tanh(x)$, so we get

$$V(x) = -2(1-\tanh^2(x)) = \frac{-2}{\cosh^2(x)}$$

the familiar soliton! This argument does *not* show this potential is reflectionless, only that it is the only candidate. However, a comparison with the formula for $\xi(x,\lambda)$ in Example 3 indeed verifies that $R_\ell(\lambda) = 0$, $\lambda > 0$ in the present case.

### 14.4. The Trace Formula for Jacobi Matrices

THEOREM 14.7. *Let $H = H_0 + V$ be a Jacobi matrix on $\mathbb{Z}^\nu$ with $V$ bounded. Let $\xi(n,\lambda)$ be $\xi_\infty$ for $A = H_0 + V$ and $\varphi = \delta_n$. Let $E_\pm = {\sup \atop \inf}\,\mathrm{spec}(H)$. Then*

$$V(n) = \tfrac{1}{2}\,(E_+ + E_-) + \int_{E_-}^{E_+} (\xi(n,\lambda) - \tfrac{1}{2})\,d\lambda \tag{14.19}$$

PROOF. (14.19) is unchanged if we add a constant, $c$, to $V$ (since then $V$, $E_+$, and $E_-$ all increase by $c$), so without loss of generality, we can suppose that $H \geq 0$. Since $\xi = 0$ (resp. 1) for $\lambda < E_-$ (resp. $\lambda > E_+$), we need only prove (14.19) for some $E_+ > \sup\mathrm{spec}(H)$ and $E_- < \inf\mathrm{spec}(H)$.

Next, we note that (14.19) can be rewritten

$$\begin{aligned} V(n) &= E_- + \int_{E_-}^{E_+} (\xi(n,\lambda) - 1)\,d\lambda \\ &= E_- + \int_{E_-}^{\infty} (\xi(n,\lambda) - 1)\,d\lambda \\ &= E_+ + \int_{-\infty}^{E_+} \xi(n,x)\,dx \end{aligned}$$

so it suffices to prove

$$V(n) = \int_0^\infty (\xi(n,x) - 1)\,dx$$

for $H \geq 0$.

But noting that

$$(\delta_n, (H+\gamma)^{-1}\delta_n) = \gamma^{-1} - \gamma^{-2}V(n) + O(\gamma^{-3})$$

this follows from (11.30) (Theorem 11.17). □

## 14.5. A Regularity Theorem for the A.C. Spectrum

In this section, we will prove the following:

THEOREM 14.8 (Gesztesy-Simon [**132**]). *Let $V_m(x)$ and $V_\infty(x)$ be functions on $\mathbb{R}$ (resp. $\mathbb{Z}$) so that*

(i) $\inf_{x\in\mathbb{R};\, m\geq 1} V_m(x) > -\infty$, *and for each* $R > 0$, $\sup_{|x|<R;\, m\geq 1}|V_m(x)| < \infty$ (resp. $\sup_{x\in\mathbb{Z};\, m\geq 1}|V_m(x)| < \infty$).

(ii) $V_m \to V_\infty$ *uniformly on compacts (resp. pointwise).*

(iii) *Each $V_m$ is periodic (but the period can be $m$-dependent).*

*Let $H_m, H_\infty$ be the Hamiltonians $-\frac{d^2}{dx^2} + V_m$ (resp. the Jacobi matrices on $\ell^2(\mathbb{Z}), h_0 + V_m$). Then (with $|\cdot|$ = Lebesgue measure):*

$$|\sigma_{\mathrm{ac}}(H_\infty)| \geq \limsup|\sigma_{\mathrm{ac}}(H_m)|$$

The most interesting application of this result is to prove a slightly strengthened version of a striking result of Last [**188**]:

COROLLARY 14.9. *Let $V(n) = \lambda\cos(\pi\alpha n + \theta)$ and let $h$ be the corresponding Jacobi matrix. Then $|\sigma_{\mathrm{ac}}(h)| \geq 4 - 2|\lambda|$.*

PROOF. If $\alpha = p/q$ is rational, this is proven in [**24**]. If $\alpha$ is arbitrary, let $p_m/q_m \to \alpha$ and let $V_m(n) = \lambda\cos(\pi_m\alpha n + \theta)$. Then $V, V_m$ obey the hypotheses of Theorem 14.8. So, by the theorem, we can take limits. □

REMARK. If $\alpha$ is a Liouville number (irrational but well approximated by rationals), it is known, for $\lambda > 2$, that $h$ has only singular continuous spectrum, and before Last [**188**], some believed that was true also for $|\lambda| < 2$. Last showed that if $|\lambda| < 2$, then $|\sigma_{\mathrm{ac}}(h)| > 0$.

To prove Theorem 14.8, we begin with

THEOREM 14.10. *Under the hypotheses* (i), (ii) *of Theorem 14.8,*

$$\xi_\infty^{(m)}(x,\lambda)\, dx \to \xi_\infty^{(\infty)}(x,\lambda)\, dx$$

*weakly as measures where $\xi_\infty^{(m)}$ is the spectral shift for $-\frac{d^2}{dx^2} + V_m$ with $\varphi = \delta(x)$.*

PROOF. It suffices to prove that for all $z \in \mathbb{C}\backslash\mathbb{R}$,

$$\int \frac{\xi_\infty^{(m)}(x,\lambda)\, d\lambda}{(\lambda - z)^2} \longrightarrow \int \frac{\xi_\infty^{(\infty)}(x,\lambda)\, d\lambda}{(\lambda - z)^2}$$

by a standard density argument. By (11.27) and the Cauchy formula, it suffices to prove that $F_m(z) \to F_\infty(z)$ where $F_m(z) = G_{V_m}(0,0;z)$. Writing

$$[(H_n - z)^{-1} - (H_\infty - z)^{-1}](H_\infty - z)\psi = (H_n - z)^{-1}(V_\infty - V_n)\psi$$

for $\psi \in C_0^\infty(\mathbb{R})$, we see that

$$\text{s-lim}(H_n - z)^{-1} = (H_\infty - z)^{-1} \tag{14.20}$$

Let $H_0 = -\frac{d^2}{dx^2}$. Then $(H_\infty + 1)^{1/2}(H_n - z)^{-1}(H_0 + 1)^{1/2}$ is uniformly bounded, so (14.20) implies that

$$\text{s-lim}(H_\infty + 1)^{1/2}(H_n - z)^{-1}(H_0 + 1)^{1/2} = (H_0 + 1)^{1/2}(H_\infty - z)^{-1}(H_0 + 1)^{1/2}$$

Since $(H_0 + 1)^{-1/2}\delta_0$ is in $L^2$, we can take matrix elements in this vector and conclude that $F_m(z) \to F_\infty(z)$ is as required. □

PROOF OF THEOREM 14.8. Since $V_m$ is periodic, $\xi_m(x,\lambda) = 0, 1$, or $\frac{1}{2}$ for a.e. $\lambda$. Fix $[a,b] \subset \mathbb{R}$ bounded and let $A_m = \{\lambda \mid \xi_m(x,\lambda) = 0\} \cap [a,b]$ for $m = 1, 2, \ldots, \infty$. Then

$$0 = \int_{A_\infty} \xi_\infty(x,\lambda)\, d\lambda = \lim \int_{A_\infty} \xi_m(x,\lambda)\, d\lambda$$

$$\geq \lim \tfrac{1}{2}\, |A_\infty \backslash A_m| \geq \tfrac{1}{2}\, (|A_\infty| - \liminf |A_m|)$$

We conclude that

$$|A_\infty| \leq \liminf |A_m|$$

A similar argument with $\xi_m = 1$ shows that

$$|\{\lambda \mid 0 < \xi_\infty(x,\lambda) < 1\} \cap [a,b]| \geq \limsup |\{\lambda \mid \xi_m(x,\lambda) = \tfrac{1}{2}\} \cap [a,b]|$$

But $\xi_\infty(x,\lambda) = \frac{1}{\pi} \arg(F_\infty(\lambda + i0))$ lies in $(0,1)$ if and only if $\operatorname{Im} F_\infty > 0$ (at points with $F_\infty(\lambda + i0)$ finite). So

$$|\sigma_{\mathrm{ac}}(H_m) \cap [a,b]| \geq |\{\lambda \mid 0 < \xi_m(x,\lambda) < 1\} \cap [a,b]|$$

with equality if $m$ is finite. □

## 14.6. Inverse Problems

These ideas play an important role in understanding inverse spectral theory. We will give an example here; see [**132**] for further examples. We will rely on the Gel'fand-Levitan theory [**123, 191**] which allows one to recover $V(x)$ from any $m_\alpha(z)$ and, in particular, from $G_{\theta=\pi/2}(0,0;z)$. In 1952, Borg [**52**] and Marčenko [**212**] proved that:

THEOREM 14.11. *Let $V(x) \to \infty$ at infinity. Let $\{E_n^{(\alpha)}\}_{n=1}^\infty$ be the eigenvalues for $-\frac{d^2}{dx^2} + V(x)$ with $\theta$ boundary conditions at $x = 0$ where $\alpha = -\cot(\theta)$. Then $V$ can be recovered from $\{E_n^{(\alpha)}\}_{n=1}^\infty$ for any two distinct values of $\alpha$.*

Actually, Marčenko [**212**] proved in addition that the two values of $\alpha$ can be recovered as well from the two sets $\{E_n^{(\alpha)}\}_{n=1}^\infty$.

We will prove a generalization of this.

THEOREM 14.12. *Suppose that $V$ is bounded below. Let $\xi_{\alpha,\beta}(\lambda)$ $(\alpha > \beta)$ be the spectral shift for going from $A_\beta$ to $A_\alpha$. Then $\alpha$, $\beta$, and $V$ can be recovered from $\xi_{\alpha,\beta}(\lambda)$.*

REMARK. This implies Theorem 14.11. For in that case, $E_n^{(\beta)} \leq E_n^{(\alpha)} \leq E_{n+1}^{(\beta)}$ and

$$\xi_{\alpha,\beta}(\lambda) = \begin{cases} 1 & E_n^{(\beta)} < \lambda < E_n^{(\alpha)} \\ 0 & E_n^{(\alpha)} < \lambda < E_{n+1}^{(\beta)} \end{cases}$$

PROOF. Let $F(z)$ be the Neumann-Green's function, $G_{\theta=\pi/2}(0,0;z)$, from which $V(x)$ can be recovered by Gel'fand-Levitan. We have that $\xi_{\alpha,\beta}(\lambda) = \xi_\alpha(\lambda) - \xi_\beta(\lambda)$ and

$$\int \frac{\xi_\alpha(\lambda)}{(\lambda - z)}\, d\lambda = \log(1 + \alpha F(z)) \qquad \alpha < \infty \tag{14.21}$$

$$\int \frac{\xi_\infty(\lambda)}{(\lambda - z)^2}\, d\lambda = \frac{F'(z)}{F(z)} \qquad \alpha = \infty \tag{14.22}$$

and

$$F(z) = (-z)^{-1/2}(1 + O(|z|)^{-1}) \tag{14.23}$$

by (11.16), (11.19), (11.27), and (14.11).

Since $\int \frac{\xi_\alpha(\lambda)}{|\lambda|+1} = \infty$ if and only if $\alpha = \infty$, we see that $\int \frac{\xi_{\alpha,\beta}(\lambda)}{|\lambda|+1}\, d\lambda = \infty$ if and only if $\alpha$ is infinite, and we can read that fact off of knowing $\xi_{\alpha,\beta}(\lambda)$. Consider first the case with $\alpha, \beta$ finite. By (14.23),

$$\begin{aligned}\log(1 + \alpha F(z)) &= \alpha F(z) - \tfrac{1}{2}\,\alpha^2 F(z)^2 + O(F(z)^3)\\ &= \alpha(-z)^{-1/2} - \tfrac{1}{2}\,\alpha^2(-z)^{-1} + O(z^{-3/2})\end{aligned}$$

Thus, by (14.21),

$$\int \frac{\xi_{\alpha,\beta}(\lambda)}{\lambda - z}\, d\lambda \sim (\alpha - \beta)(-z)^{-1/2} - \tfrac{1}{2}\,(\alpha^2 - \beta^2)(-z)^{-1} + O(z^{-3/2})$$

so we can find $\alpha - \beta$ and $\alpha^2 - \beta^2$, and so $\alpha$ and $\beta$ from $\xi_{\alpha,\beta}$. Moreover, since

$$\int \frac{\xi_{\alpha,\beta}(\lambda)}{\lambda - z}\, d\lambda = \log\left(\frac{1 + \alpha F(z)}{1 + \beta F(z)}\right)$$

$\xi_{\alpha,\beta}$ determines $F(z)$ and so $V$ by Gel'fand-Levitan.

Now consider the case $\alpha = \infty$. Then, by (14.23) and a contour integral,

$$\begin{aligned}\frac{F'(z)}{F(z)} &= -\frac{1}{2}\,(-z)^{-1} + O(z^{-2})\\ \frac{\beta F'(z)}{1 + \beta F(z)} &= -\frac{\beta}{2}\,(-z)^{-3/2} + O(z^{-2})\end{aligned}$$

Thus, by (14.22) and the discussion of (14.21),

$$\begin{aligned}\int \frac{[\xi_\infty(\lambda) - \xi_\beta(\lambda)]}{(x - z)^2}\, d\lambda &= \frac{\beta F'(z)}{(1 + \beta F(z)} - \frac{F'(z)}{F(z)}\\ &= \frac{1}{2}\,(-z)^{-1} - \frac{\beta}{2}\,(-z)^{-3/2} + O(z^{-2})\end{aligned}$$

so we can read $\beta$ off of $\xi_{\infty,\beta}(\lambda)$. By the above,

$$\begin{aligned}\int \frac{\xi_{\infty,\beta}(\lambda)}{(\lambda - z)^2} &= \frac{d}{dz}\,\log(\beta + F(z)^{-1})\\ &= -\frac{1}{2}\,z^{-1} + \frac{d}{dz}\,\log((-z)^{1/2}[F(z)^{-1} + \beta])\\ &\equiv -\frac{1}{2}\,z^{-1} + \frac{d}{dz}\,[H_\beta(z)]\end{aligned}$$

By the above, $H_\beta(-\gamma) \to 0$ as $\gamma \to \infty$. Thus

$$H_\beta(z) = \lim_{\gamma\to\infty} \int \frac{(\xi_{\infty,\beta} - \frac{1}{2})(\lambda)}{\lambda - z}\left(\frac{\gamma}{\gamma + \lambda}\right) d\lambda$$

and we can recover $H_\beta$ and so $F$ from $\xi_{\infty,\beta}(\lambda)$. Gel'fand-Levitan completes the proof. □

# Addenda

Our purpose here is to discuss some of the developments since the publication of the first edition, focusing especially on issues discussed there.

### A. Rearrangement Inequalities (Section 1.4)

There is an enormous literature on this subject, from which I picked out only the approach of Marcus (Theorem 1.9). The heroes of the history include Hardy, Littlewood, Schur, and Birkhoff. The book of Marshall and Olkin [**215**] is devoted to this subject, with detailed history, pictures, and more.

### B. Spectral Averaging and the Krein Spectral Shift (Sections 11.3 and 11.4)

Birman-Solomjak [**43**] have a lovely extension of (11.23) which says

$$\int_\beta^\gamma (d\mu_\alpha(E))\,d\alpha = (\xi_\gamma(E) - \xi_\beta(E))\,dE \tag{A.1}$$

where $\xi_\beta$ is the spectral shift from $A$ to $A + \beta(\varphi,\,\cdot\,\varphi)$ and $d\mu_\alpha$ is defined by

$$(\varphi, \exp[it(A + \alpha(\varphi,\,\cdot\,)\varphi)]\varphi) = \int e^{itx}\,d\mu_\alpha(x) \tag{A.2}$$

The extension is:

THEOREM A.1. *Let $A(\alpha)$, $\alpha_0 \le \alpha \le \alpha_1$, be a family of bounded self-adjoint operators so that $B(\alpha) = dA/d\alpha$ exists in weak sense. Suppose $B(\alpha)$ is trace class and positive, and $\alpha \mapsto B(\alpha)$ is continuous in trace norm. Define $d\mu_\alpha$ by*

$$\operatorname{Tr}(B(\alpha)\exp(it(A(\alpha)))) = \int e^{itx}\,d\mu_\alpha(x) \tag{A.3}$$

*and let $\xi_\beta$ be the spectral shift from $A(\alpha_0)$ to $A(\beta)$. Then* (A.1) *holds.*

A simple proof of this result is due to Simon [**309**], who derives it from

$$\frac{d}{ds}\operatorname{Tr}(f(A(s))) = \operatorname{Tr}(A'(s)f'(A(s))) \tag{A.4}$$

for suitable $C^1$ functions $f$.

### C. Löwner's Theorem and Segal's Inequality (Section 8.1)

There is an interesting relation between Segal's inequality (8.3) that ($\|\cdot\|$ = operator norm)

$$\|e^{A+B}\| \le \|e^{A/2}e^Be^{A/2}\| \tag{A.5}$$

and Löwner's theorem [**210, 103**] that

$$0 \le C \le D \Rightarrow \log C \le \log D \tag{A.6}$$

Let us show that (A.6) implies (A.5) for finite matrices.

Let $c = \|e^{A/2}e^{B}e^{A/2}\|$. Since $e^{A/2}e^{B}e^{A/2}$ is Hermitian,

$$e^{A/2}e^{B}e^{A/2} \le c \Rightarrow e^{B} \le c\,e^{-A} \tag{A.7}$$

Using (A.6),

$$B \le \log c - A \Rightarrow A + B \le \log c \tag{A.8}$$

Since $\log c$ is a number, this implies, by the spectral theorem, that $e^{A+B} \le c$, which implies $\|e^{A+B}\| \le c$, which is (A.5).

## D. Brislawn's Theorem (Chapters 2 and 3)

Brislawn [**57, 58, 59**] found far-reaching generalizations of Theorem 2.12. While he has results for general measure spaces [**59**], let us describe his theorems for operators on $\mathbb{R}^\nu$.

Given a function $f \in L^1_{\text{loc}}(\mathbb{R}^k)$, we define its maximal function by

$$f^*(y) = \sup_r \, (A_r|f|)(y)$$

where $(A_r f)(y) = |C_r|^{-1} \int_{C_r} f(x+y)\,dx$ with $C_r$ the hypercube of side $r$. For each $L^1_{\text{loc}}$-function, the Lebesgue value $\tilde{f}(y) \equiv \lim_{r\downarrow 0}(A_r f)(y)$ exists for a.e. $y$. Note that $|\tilde{f}| \le f^*$. Here is Brislawn's first theorem:

THEOREM A.2. *Let $A$ be a trace class operator on $L^2(\mathbb{R}^\nu)$ and let $K(x,y)$ be its integral kernel. Let $K^*(x,y)$ be the $k = 2\nu$ maximal function and $\widetilde{K}(x,y)$ its Lebesgue value. Then*

(a) $$\int K^*(x,x)\,d^\nu x < \infty \tag{A.9}$$

(b) $\widetilde{K}(x,x)$ *exists for a.e.* $x$

(c) $$\operatorname{Tr}(A) = \int \widetilde{K}(x,x)\,d^\nu x \tag{A.10}$$

REMARKS. 1. Since $A$ is trace class, it is Hilbert-Schmidt and so has an integral kernel $K(x,y)$.

2. We emphasize $K^*$ and $\widetilde{K}$ are defined via $2n$-dimensional averaging, so are independent of measure zero uncertainties in $K$.

3. Lebesgue's theorem only says that $\widetilde{K}(x,y)$ exist for a.e. $x, y$, and so (b) goes beyond Lebesgue's theorem by making an assertion about the diagonal.

For positive operators, Brislawn has a converse. We emphasize again that $K^*$ is a $2n$-dimensional maximal function.

THEOREM A.3. *Let $A$ be a positive Hilbert-Schmidt operator on $\mathbb{R}^\nu$ and let $K(x,y)$ be its integral kernel. Then $A$ is trace class if and only if* (A.9) *holds.*

This theorem relies heavily on earlier work of Weidmann [**338**]. For related results, see Lorentz-Rejto [**208**].

## E. Pointwise Domination (Chapter 2)

The conjecture after Theorem 2.13 is false! This was first shown by Peller [**234**] who determined necessary and sufficient conditions for a Hankel operator to be in $\mathcal{I}_p$, and was therefore able to construct counterexamples; see also Peller [**235**].

Lon Rosen [**258**] looked at when the conjecture holds for finite matrices. His result is that if $p$ is not an even integer, then $0 \le b_{ij} \le a_{ij}$ implies $\|B\|_p \le \|A\|_p$ if and only if $n < 1 + \frac{1}{2}p$. For an earlier result of this genre, see Dechamps-Gondim et al. [**85**].

## F. Analytic Continuation of the One-Dimensional Birman-Schwinger Kernel (Chapter 5)

Proposition 5.6 has been enhanced to study resonances. In that case, Froese [**116**] and Simon [**310**] have proven

THEOREM A.4. *Suppose $V$ has compact support on $\mathbb{R}$. Then the kernel $K$ defines an entire trace class valued function of $k$.*

*If $V$ decays exponentially, say, $|V(x)| \le C\,e^{-A|x|}$, then the kernel $K$ is trace class valued in $\{k \mid \operatorname{Im} k > -A\}$.*

## G. Trace Class Decoupling of Jacobi and Other Matrices (Chapter 6 and Section 11.7)

Classical Jacobi matrices are slightly different from what we described in Section 11.7, that is, the discrete Schrödinger operators. $J$ is associated to a pair of sequences of reals $\{a_n\}_{n=1}^\infty$ and $\{b_n\}_{n=1}^\infty$ with $a_n > 0$. $J$ is the matrix

$$J = \begin{pmatrix} b_1 & a_1 & 0 & \dots \\ a_1 & b_2 & a_2 & \dots \\ 0 & a_2 & b_3 & \dots \\ \dots & \dots & \dots & \dots \\ \dots & \dots & \dots & \dots \end{pmatrix} \tag{A.11}$$

This is on $\mathbb{Z}_+$ rather than $\mathbb{Z}^\nu$ and the $a_n$'s need not be one.

Dombrowski [**99, 100, 101**] used trace class scattering to prove:

THEOREM A.5 (Dombrowski). *Let $J$ be a Jacobi matrix obeying* $\sup_n(|a_n| + |b_n|) < \infty$ *and*

$$\liminf_{n\to\infty} |a_n| = 0 \tag{A.12}$$

*Then the spectral measure for $J$ has no absolutely continuous part.*

PROOF. Pick a subsequence $n(j)$ with

$$\sum_{j=1}^\infty a_{n(j)} < \infty \tag{A.13}$$

Let $\tilde J$ be the Jacobi matrix with each $a_{n(j)}$ set to 0. By (A.13), $J - \tilde J$ is trace class, so by Corollary 6.5, the a.c. parts of $J$ and $\tilde J$ are unitarily equivalent. But $\tilde J$ is a direct sum of finite matrices, and so it has pure point spectrum and so no a.c. part. □

Unaware of Dombrowski's work, Simon-Spencer [**313**] found a similar argument to control the case $a_j \equiv 1$ and $\limsup |b_j| = \infty$ and suitable one-dimensional Schrödinger operators. Golinskii-Simon (see Section 4.3 of [**311**]) proved a variant of this result applicable to orthogonal polynomials on the unit circle.

## H. Kadec-Klee Property (Chapter 2)

After Theorem 2.21, we asked if the theorem remained true for $\mathcal{J}_1$. This was settled by Arazy [**14**], who proved a more general result:

THEOREM A.6 (Arazy). *Suppose a symmetric norm* $\Phi$ *on sequences has the property*

$$x_j^{(n)} \to x_j^{(\infty)} \text{ all } j + \Phi(x^{(n)}) \to \Phi(x^{(\infty)}) \Rightarrow \Phi(x^{(n)} - x^{(\infty)}) \to 0 \tag{A.14}$$

*Then the corresponding symmetrically normed ideals obey*

$$A^{(n)} \xrightarrow{w} A^{(\infty)} + \Phi(A^{(n)}) \to \Phi(A^{(\infty)}) \Rightarrow \Phi(A^{(n)} - A^{(\infty)}) \to 0 \tag{A.15}$$

REMARKS. 1. It is easy to see that $\ell^1$ has the property (A.14), so Theorem 2.21 does hold for $\mathcal{J}_1$.

2. These results are closely connected to the Kadec-Klee property for Banach spaces. This property says:

$$x_n \xrightarrow{w} x_\infty + \|x_n\| \to \|x_\infty\| \Rightarrow \|x_n - x_\infty\| \to 0 \tag{A.16}$$

Note that in (A.16), $\xrightarrow{w}$ means weak Banach space convergence, while in (A.15) it means weak operator convergence.

Simon [**304**] has a simple proof of Theorem A.6. See Dilworth-Hsu [**94**] and Dodds et al. [**70, 97**] for further work on the subject. Arazy [**13, 15, 16, 17, 18**] has additional interesting work on symmetrically normed operator ideals.

## I. Rank One Perturbations of Unitaries (Section 11.2)

Parallel to the theory of rank one perturbations of self-adjoint operators is the theory of rank one perturbations of unitaries which have the form

$$V = U + (\lambda - 1)(\varphi, \cdot\,)U\varphi \tag{A.17}$$

with $\lambda = e^{i\theta}$ in $\{z \mid |z| = 1\}$. Equivalently,

$$V\psi = U\psi \quad \text{if} \quad \psi \perp \varphi \tag{A.18}$$

$$V\varphi = \lambda(U\varphi) \tag{A.19}$$

The formulae for this setting are discussed in Subsections 1.3.9 and 1.4.16 of [**311**].

There are more general rank $n$ perturbation formulae discovered in Section 4.5 of [**311**]. Here is a general result: Let $P$ be the orthogonal projection onto a finite-dimensional space. Let $\Lambda : \operatorname{Ran} P \to \operatorname{Ran} P$ be unitary, and suppose $U$ is unitary and

$$V = U(1 - P) + U\Lambda P \tag{A.20}$$

For $|z| < 1$, define $G(z)$, $G_0(z)$, $g(z)$, and $g_0(z)$ as operators from $\operatorname{Ran} P$ to itself by

$$G(z) = P\left[\frac{V+z}{V-z}\right]P = (1 + zg(z))(1 - zg(z))^{-1} \tag{A.21}$$

$$G_0(z) = P\left[\frac{U+z}{U-z}\right]P = (1 + zg_0(z))(1 - zg(z))^{-1} \tag{A.22}$$

THEOREM A.7 (= Theorem 4.5.6 of [**311**]). *Under* (A.20)–(A.22),

$$g = \Lambda^{-1} g_0 \tag{A.23}$$

## J. Theorem of de la Vallée-Poussin (Theorem 11.6(ii))

At the time I wrote the material in Chapter 11, the only proof I knew of this is the one in Saks, which is quite complicated. I have since realized that the result is quite simple and that simpler proofs are in many places in the literature; for example, Theorem 7.15 of Rudin [**261**] or Jakšić [**155**].

## K. Poltoratskii's Theorem (Sections 11.1 and 11.2)

If we make the $\mu$-dependence of (11.5) explicit by referring to $F_\mu(z)$, then the same ideas that proved (11.8) prove

THEOREM A.8. *Let $f \in L^1(d\mu)$. Then for $\mu$ a.e. $x$,*

$$\lim_{\varepsilon \downarrow 0} \frac{\operatorname{Im} F_{f\mu}(x+i\varepsilon)}{\operatorname{Im} F_\mu(x+i\varepsilon)} = f(x) \tag{A.24}$$

Much more subtle is the following

THEOREM A.9 (Poltoratskii [**242**]). *Let $f \in L^1(d\mu)$. Then for $\mu_{\text{sing}}$ a.e. $x$,*

$$\lim_{\varepsilon \downarrow 0} \frac{F_{f\mu}(x+i\varepsilon)}{F_\mu(x+i\varepsilon)} = f(x) \tag{A.25}$$

REMARKS. 1. Here $\mu_{\text{sing}}$ means singular with respect to Lebesgue measure.

2. It is easy to see this fails for a.c. parts and it follows from Theorem 11.6(iii) for the pure points of $d\mu$. The subtlety is the s.c. part of $d\mu$.

We mention this theorem here because Jakšić-Last [**158**] have a simple proof of Theorem A.9 using the theory of rank one perturbations; see also [**159, 155**].

## L. Cyclicity and Related Issues in the Anderson Model (Chapter 13)

In [**307**], Simon proved:

THEOREM A.10. *Let $\{V_\omega(n)\}_{n\in\mathbb{Z}^\nu}$ be independent, identically distributed random variables with a density that is absolutely continuous. Let $h_\omega$ be the operator on $\ell^2(\mathbb{Z}^\nu)$ given by*

$$(h_\omega u)(n) = \sum_{|m|=1} u(m+n) + V_\omega(n)u(n) \tag{A.26}$$

*Suppose that for a.e. $\omega$, $\operatorname{spec}(h_\omega)$ is pure point on some interval $[a,b]$. Then for a.e. $\omega$ and all $n$, each $\delta_n$ is cyclic for $h_\omega P_{[a,b]}(h_\omega)$. In particular, the spectrum is simple there.*

The proof uses rank one perturbation theory. By exploiting the ideas in this proof, Jakšić-Last [**156, 159**] have proven several deep results about operators of the form (A.26).

## M. $L^p$ Bounds on the Spectral Shift (Section 11.4)

One has two simple bounds on the Krein spectral shift:

$$\int |\xi_{C,A}(x)|\,dx \le \operatorname{Tr}(|C-A|) \tag{A.27}$$

$$\sup_x |\xi_{C,A}(x)| \le \operatorname{rank}(C-A) \tag{A.28}$$

The first is (11.26b) and the second follows from Theorem 11.9. Combes, Hislop, and Nakamura [**77**] have proven the following:

THEOREM A.11 (Combes-Hislop-Nakamura [**77**]). *For $1 \leq p < \infty$, we have*

$$\left( \int |\xi_{C,A}(x)|^p \, dx \right)^{1/p} \leq \sum_n \mu_n (C-A)^{1/p} \tag{A.29}$$

Notice that as $p \to \infty$, the right side of (A.29) converges to $\operatorname{rank}(C-A)$. In this sense, (A.29) interpolates between (A.27) and (A.28). Hundertmark-Simon [**152**] have proven a stronger result:

$$\left( \int |\xi_{C,A}(x)|^p \, dx \right) \leq \sum_{n=1}^{\infty} [n^p - (n-1)^p] \mu_n (C-A) \tag{A.30}$$

It is not immediately obvious, but as proven by [**152**], for $\mu_1 \geq \mu_2 \geq \cdots \geq 0$ and $1 \leq p < \infty$,

$$\left( \sum_{n=1}^{\infty} [n^p - (n-1)^p] \mu_n \right)^{1/p} \leq \sum_{n=1}^{\infty} \mu_n^{1/p} \tag{A.31}$$

so (A.30) implies (A.29).

## N. Trace of Trace Class Projections (Chapter 3)

The following seems obvious until you think about how to prove it!

THEOREM A.12 (Effros [**108**]). *If $P$ and $Q$ are Hermitian projections so that $P - Q \in \mathcal{I}_1$, then $\operatorname{Tr}(P-Q)$ is an integer.*

Avron, Seiler, and Simon [**22**] found an interesting way of understanding and proving this result using

$$A = P - Q \qquad B = 1 - P - Q \tag{A.32}$$

Using $P^2 = P$ and $Q^2 = Q$, one finds

$$A^2 + B^2 = 1 \qquad AB + BA = 0 \tag{A.33}$$

If $A$ is trace class, then it has a complete set of eigenvectors. Let $0 < \lambda < 1$ and let $\mathcal{H}_{\pm\lambda}$ be the set of vectors with $A\varphi = \pm\lambda\varphi$. If $\varphi \in \mathcal{H}_\lambda$, then by (A.33),

$$A(B\varphi) = -\lambda B\varphi \qquad B^2\varphi = (1-\lambda^2)\varphi \tag{A.34}$$

The first formula says $B : \mathcal{H}_\lambda \to \mathcal{H}_{-\lambda}$, and the second says $B$ is invertible from $\mathcal{H}_\lambda$ to $\mathcal{H}_{-\lambda}$. It follows that

$$n_\lambda \equiv \dim(\mathcal{H}_\lambda) = n_{-\lambda} \tag{A.35}$$

Thus

$$\operatorname{Tr}(A) = \sum_\lambda n_\lambda \lambda = n_1 - n_{-1} \tag{A.36}$$

is an integer. See more on this set-up in [**22, 83, 322, 325**].

## O. The Xi Function and Borg's Theorem (Chapter 14)

In [**51**], Borg proved a celebrated result that if

$$H = -\frac{d^2}{dx^2} + V \tag{A.37}$$

on $L^2(\mathbb{R})$ with $V$ periodic, that is, $V(x+L) = V(x)$ for some $L$, and if $\operatorname{spec}(H) = [0, \infty)$ (i.e., no gaps in the spectrum), then $V = 0$. Clark et al. [**73**] realized that the xi function is an ideal tool for understanding and extending Borg's theorem.

It is easy to describe their idea in the case of periodic whole-line discrete Schrödinger matrices. As explained in Example 1 of Section 14.3, in the periodic case, $G(n,n,E+i0)$ is real on the spectrum of $h$, so $\xi(n,\lambda)=\frac{1}{2}$ on $\operatorname{spec}(h)$. If $\operatorname{spec}(h)=[E_-,E_+]$ (i.e., no gaps), then (14.19) implies that $V(n)=\frac{1}{2}(E_-+E_+)$, so $V(n)$ is a constant!

This idea can be used for reflectionless potentials (any potential where $G(n,n,E+i0)$ is real on the spectrum), suitable matrix-valued Schrödinger operators, and suitable Dirac operators; see [**28, 72, 73**]. Gesztesy-Zinchenko [**133**] have extended the idea to orthogonal polynomials on the unit circle.

## P. Further Developments of the Aizenman-Molchanov Theory (Chapter 13)

There have been many developments in the theory of localization using fractional moments, a theory invented by Aizenman-Molchanov, whose ideas are described in Chapter 13.

Aizenman [**4**] proved a theorem that goes from bounds of the form

$$\int_a^b \mathbb{E}(|(\delta_n,(h_\omega-\lambda-i0)^{-1}\delta_m)|^p)\leq C_1 e^{-\kappa_1|n-m|} \tag{A.38}$$

for some $0<p<1$ to bounds of the form

$$\mathbb{E}\big(\sup_t |(\delta_n, P_{[a,b]}(h_\omega)e^{-ith_\omega}\delta_m)|\big)\leq C_2 e^{-\kappa_2|n-m|} \tag{A.39}$$

Del Rio et al. [**90**] rearranged this proof and used it to prove a strong form (SULE) of localization of eigenfunctions. These ideas were exploited by Minami [**222**] to prove Poisson statistics for eigenvalues in a large box.

Further developments can be found in Aizenman-Graf [**7**], Aizenman et al. [**9, 10**], and Jakšić et al. [**157, 160**]. Extensions to continuum Schrödinger operators are due to Aizenman et al. [**5, 6**]; extensions to orthogonal polynomials on the unit circle are due to Stoiciu [**321**]. For the latter case, Simon [**312**] found an analog of (A.38) $\Rightarrow$ (A.39).

## Q. Further Results on the Krein Spectral Shift (Section 11.4)

There is an enormous literature on the Krein spectral shift and its applications. For extensions to non-trace class situations, see Dostanić [**104**], Koplienko [**172**], Neidhardt [**227**], and Sadovnichii-Podol'skii [**273, 274**]. Without trying to be comprehensive, here is a list of some papers on the general theory: Adamjan-Langer [**1**], Adamjan-Neidhardt [**2**], Albeverio-Makarov-Motovilov [**11**], Birman-Pushnitskii [**36**], Birman-Yafaev [**46**], Bollé et al. [**50**], Bouclet [**53**], Boyadzhiev [**54**], Bruneau-Petkov [**60, 61**], Dimassi-Petkov [**95**], Geisler-Kostrykin-Schrader [**122**], Gesztesy-Makarov [**125**], Gesztesy-Makarov-Motovilov [**126**], Gesztesy-Makarov-Naboko [**127**], Hundertmark et al. [**151**], Ivrii-Shubin [**154**], Kostrykin [**173**], Kostrykin-Schrader [**174, 175**], Kreĭn-Yavryan [**179**], Langer-de Snoo-Yavryan [**187**], Mohapatra-Sinha [**224, 315**], Nakamura [**226**], Poltoratskii [**243**], Pushnitskii [**244, 245, 246, 247**], Pushnitskii-Ruzhanskii [**248, 249**], Robert [**257**], Rybkin [**271, 272**], Safronov [**275**], Shirai [**290**], Sobolev [**318**], Vasy-Wang [**335**], and Yavryan [**343**]. For applications to magnetic Schrödinger operators, see Bruneau-Pushnitskii-Raikov [**62, 63**] and Fernández-Raikov [**113**].

### R. More on Convex and Concave Trace Functions (Theorem 8.10)

There has been considerable literature studying results connected to Lieb concavity. In particular, Epstein [**111**] proved that if $B > 0$ and $0 < p < 1$, then $A \mapsto \mathrm{Tr}((BA^pB)^{1/p})$ is concave in $A$ when considered on the positive operators. For further results along these lines, see Bekjan [**27**], Bhatia [**31, 32**], Carlen-Lieb [**69**], Hiai [**148**], and Ruskai [**264**].

### S. Further Results on Uniform Convexity (Section 1.9)

For more recent discussion of the inequalities discussed in Section 1.9, see Hanner [**146**], Tomczak-Jaegermann [**331**], Pisier [**238**], and Ball-Carlen-Lieb [**25**].

# Bibliography


[1] V. M. Adamjan and H. Langer, *The spectral shift function for certain block operator matrices*, Math. Nachr. **211** (2000), 5–24.

[2] V. M. Adamjan and H. Neidhardt, *On the summability of the spectral shift function for pair of contractions and dissipative operators*, J. Operator Theory **24** (1990), 187–205.

[3] S. Agmon, *Spectral properties of Schrödinger operators and scattering theory*, Ann. Scuola Norm. Sup. Pisa Cl. Sci. (4) **2** (1975), 151–218.

[4] M. Aizenman, *Localization at weak disorder: Some elementary bounds*, Rev. Math. Phys. **6** (1994), 1163–1182.

[5] M. Aizenman, A. Elgart, S. Naboko, J. H. Schenker, and G. Stolz, *Fractional moment methods for Anderson localization in the continuum*, to appear in Proc. Int. Cong. Math. Phys. (Lisbon, 2003), World Scientific, Singapore.

[6] M. Aizenman, A. Elgart, S. Naboko, J. H. Schenker, and G. Stolz, *Moment analysis for localization in random Schrödinger operators*, preprint.

[7] M. Aizenman and G. M. Graf, *Localization bounds for an electron gas*, J. Phys. A **31** (1998), 6783–6806.

[8] M. Aizenman and S. Molchanov, *Localization at large disorder and at extreme energies: An elementary derivation*, Comm. Math. Phys. **157** (1993), 245–278.

[9] M. Aizenman, J. H. Schenker, R. M. Friedrich, and D. Hundertmark, *Constructive fractional-moment criteria for localization in random operators*, Statistical Mechanics: From Rigorous Results to Applications, Phys. A **279** (2000), 369–377.

[10] M. Aizenman, J. H. Schenker, R. M. Friedrich, and D. Hundertmark, *Finite-volume fractional-moment criteria for Anderson localization*, Dedicated to Joel L. Lebowitz, Comm. Math. Phys. **224** (2001), 219–253.

[11] S. Albeverio, K. A. Makarov, and A. K. Motovilov, *Graph subspaces and the spectral shift function*, Canad. J. Math. **55** (2003), 449–503.

[12] T. Ando, *Topics on Operator Inequalities*, Division of Applied Mathematics, Research Institute of Applied Electricity, Hokkaido University, Sapporo, 1978.

[13] J. Arazy, *Basic sequences, embeddings, and the uniqueness of the symmetric structure in unitary matrix spaces*, J. Funct. Anal. **40** (1981), 302–340.

[14] J. Arazy, *More on convergence in unitary matrix spaces*, Proc. Amer. Math. Soc. **83** (1981), 44–48.

[15] J. Arazy, *On the geometry of the unit ball of unitary matrix spaces*, Integral Equations Operator Theory **4** (1981), 151–171.

[16] J. Arazy, *On stability of unitary matrix spaces*, Proc. Amer. Math. Soc. **87** (1983), 317–321.

[17] J. Arazy, *Isomorphisms of unitary matrix spaces*, Banach Space Theory and Its Applications (Bucharest, 1981), pp. 1–6, Lecture Notes in Math., 991, Springer-Verlag, Berlin-New York, 1983.

[18] J. Arazy and P.-K. Lin, *On p-convexity and q-concavity of unitary matrix spaces*, Integral Equations Operator Theory **8** (1985), 295–313.

[19] N. Aronszajn, *On a problem of Weyl in the theory of singular Sturm-Liouville equations*, Amer. J. Math. **79** (1957), 597–610.

[20] N. Aronszajn and W. F. Donoghue, *On exponential representations of analytic functions in the upper half-plane with positive imaginary part*, J. Analyse Math. **5** (1957), 321–388.

[21] J. Avron, I. Herbst, and B. Simon, *Schrödinger operators with magnetic fields. III. Atoms in homogeneous magnetic field*, Comm. Math. Phys. **79** (1981), 529–572.

[22] J. Avron, R. Seiler, and B. Simon, *The index of a pair of projections*, J. Funct. Anal. **120** (1994), 220–237.

[23] J. Avron and B. Simon, *Almost periodic Schrödinger operators. II. The integrated density of states*, Duke Math. J. **50** (1983), 369–391.

[24] J. Avron, B. Simon, and P. H. M. van Mouche, *On the measure of the spectrum for the almost Mathieu operator*, Comm. Math. Phys. **132** (1990), 103–118.

[25] K. Ball, E. A. Carlen, and E. H. Lieb, *Sharp uniform convexity and smoothness inequalities for trace norms*, Invent. Math. **115** (1994), 463–482.

[26] V. Bargmann, *On the number of bound states in a central field of force*, Proc. Nat. Acad. Sci. U.S.A. **38** (1952), 961–966.

[27] T. N. Bekjan, *On joint convexity of trace functions*, Linear Algebra Appl. **390** (2004), 321–327.

[28] E. D. Belokolos, F. Gesztesy, K. A. Makarov, and L. A. Sakhnovich, *Matrix-valued generalizations of the theorems of Borg and Hochstadt*, Evolution Equations, pp.1–34, Lecture Notes in Pure and Appl. Math., 234, Dekker, New York, 2003.

[29] G. Bennett, *Some ideals of operators on Hilbert space*, Studia Math. **55** (1975/76), 27–40.

[30] F. A. Berezin, *Convex functions of operators*, Mat. Sb. (N.S.) **88(130)** (1972), 268–276.

[31] R. Bhatia, *Matrix Analysis*, Graduate Texts in Mathematics, 169, Springer-Verlag, New York, 1997.

[32] R. Bhatia, *Partial traces and entropy inequalities*, Linear Algebra Appl. **370** (2003), 125–132.

[33] G. Birkhoff, *Three observations on linear algebra*, Univ. Nac. Tucumán. Revista A. **5** (1946), 147–151.

[34] M. Š. Birman, *On the number of eigenvalues in a quantum scattering problem*, Vestnik Leningrad. Univ. **16** (1961), no. 13, 163–166 [Russian].

[35] M. Š. Birman and V. V. Borzov, *On the asymptotic formula for the discrete spectrum of certain singular differential operators*, Topics in Mathematical Physics, Vol. 5: Spectral Theory, pp. 19–30. Consultants Bureau, New York-London, 1972.

[36] M. Š. Birman and A. B. Pushnitskii, *Spectral shift function, amazing and multifaceted*, Dedicated to the memory of Mark Grigorievich Krein (1907–1989), Integral Equations Operator Theory **30** (1998), 191–199.

[37] M. Š. Birman and M. Z. Solomjak, *Estimates of singular numbers of integral operators. I*, Vestnik Leningrad. Univ. **22** (1967), no. 7, 43–53 [Russian].

[38] M. Š. Birman and M. Z. Solomjak, *Estimates of singular numbers of integral operators. II*, Vestnik Leningrad. Univ. **22** (1967), no. 13, 21–28 [Russian].

[39] M. Š. Birman and M. Z. Solomjak, *Piecewise-polynomial approximations of functions of classes* ${W_p}^\alpha$, Math. USSR Sb. **2** (1967), 295–317; Russian original in Mat. Sb. (N.S.) **73 (115)** (1967), 331–355.

[40] M. Š. Birman and M. Z. Solomjak, *Estimates of singular numbers of integral operators. III. Operators in unbounded domains*, Vestnik Leningrad Univ. Math. **2** (1975), 9–27; Russian original in Vestnik Leningrad. Univ. **24** (1969), 35–48.

[41] M. Š. Birman and M. Z. Solomjak, *Spectral asymptotics of nonsmooth elliptic operators. I*, Trans. Moscow Math. Soc. **27** (1972), 1–52 (1975); Russian original in Trudy Moskov. Mat. Obšč. **27** (1972), 3–52.

[42] M. Š. Birman and M. Z. Solomjak, *Spectral asymptotics of nonsmooth elliptic operators. II*, Trans. Moscow Math. Soc. **28** (1973), 1–32 (1975); Russian original in Trudy Moskov. Mat. Obšč. **28** (1973), 3–34.

[43] M. Š. Birman and M. Z. Solomjak, *Remarks on the spectral shift function*, J. Soviet Math. **3** (1975), 408–419.

[44] M. Š. Birman and M. Z. Solomjak, *Estimates of singular numbers of integral operators*, Russian Math. Surveys **32** (1977), 15–89; Russian original in Uspekhi Mat. Nauk **32** (1977), 17–84, 271.

[45] M. Š. Birman and M. Z. Solomjak, *Quantitative analysis in Sobolev imbedding theorems and applications to spectral theory*, American Mathematical Society Translations, Series 2, vol. 114, American Mathematical Society, Providence, RI, 1980.

[46] M. Š. Birman and D. R. Yafaev, *The spectral shift function. The papers of M. G. Krein and their further development*, St. Petersburg Math. J. **4** (1993), 833–870; Russian original in Algebra i Analiz **4** (1992), 1–44.

[47] R. Blankenbecler, M. L. Goldberger, and B. Simon, *The bound states of weakly coupled long-range one-dimensional quantum Hamiltonians*, Ann. Physics **108** (1977), 69–78.

[48] R. P. Boas, *Entire Functions*, Academic Press, New York, 1954.

[49] R. P. Boas, *Some uniformly convex spaces*, Bull. Amer. Math. Soc. **46** (1940), 304–311.

[50] D. Bollé, F. Gesztesy, H. Grosse, and B. Simon, *Krein's spectral shift function and Fredholm determinants as efficient methods to study supersymmetric quantum mechanics*, Lett. Math. Phys. **13** (1987), 127–133.

[51] G. Borg, *Eine Umkehrung der Sturm-Liouvilleschen Eigenwertaufgabe. Bestimmung der Differentialgleichung durch die Eigenwerte*, Acta Math. **78** (1946), 1–96.

[52] G. Borg, *Uniqueness theorems in the spectral theory of* $y'' + (\lambda - q(x))y = 0$, Proc. 11th Scandinavian Congress of Mathematicians (Trondheim, 1949), pp. 276–287, Johan Grundt Tanums Forlag, Oslo, 1952.

[53] J.-M. Bouclet, *Spectral distributions for long range perturbations*, J. Funct. Anal. **212** (2004), 431–471.

[54] K. N. Boyadzhiev, *Krein's trace formula and the spectral shift function*, Int. J. Math. Math. Sci. **25** (2001), 239–252.

[55] H. J. Brascamp, *The Fredholm theory of integral equations for special types of compact operators on a separable Hilbert space*, Compos. Math. **21** (1969), 59–80.

[56] H. J. Brascamp, E. H. Lieb, and J. M. Luttinger, *A general rearrangement inequality for multiple integrals*, J. Funct. Anal. **17** (1974), 227–237.

[57] C. Brislawn, *Kernels of trace class operators*, Proc. Amer. Math. Soc. **104** (1988), 1181–1190.

[58] C. Brislawn, *Trace class integral kernels*, Operator theory: Operator Algebras and Applications, Part 2 (Durham, NH, 1988), pp. 61–64, Proc. Sympos. Pure Math., 51, Part 2, American Mathematical Society, Providence, RI, 1990.

[59] C. Brislawn, *Traceable integral kernels on countably generated measure spaces*, Pacific J. Math. **150** (1991), 229–240.

[60] V. Bruneau and V. Petkov, *Representation of the spectral shift function and spectral asymptotics for trapping perturbations*, Comm. Partial Differential Equations **26** (2001), 2081–2119.

[61] V. Bruneau and V. Petkov, *Meromorphic continuation of the spectral shift function*, Duke Math. J. **116** (2003), 389–430.

[62] V. Bruneau, A. Pushnitskii, and G. Raikov, *Spectral shift function in strong magnetic fields*, Algebra i Analiz **16** (2004), 207–238.

[63] V. Bruneau and G. Raikov, *High energy asymptotics of the magnetic spectral shift function*, J. Math. Phys. **45** (2004), 3453–3461.

[64] D. C. Brydges, J. Fröhlich, and E. Seiler, *On the construction of quantized gauge fields. III. The two-dimensional abelian Higgs model without cutoffs*, Comm. Math. Phys. **79** (1981), 353–399.

[65] A.-P. Calderón, *Intermediate spaces and interpolation, the complex method*, Studia Math. **24** (1964), 113–190.

[66] J. W. Calkin, *Two-sided ideals and congruences in the ring of bounded operators in Hilbert space*, Ann. of Math. (2) **42** (1941), 839–873.

[67] T. Carleman, *Über die Fourierkoeffizienten einer stetigen Funktion*, Acta Math. **41** (1918), 377–384.

[68] T. Carleman, *Zur Theorie der linearen Integralgleichungen*, Math. Z. **9** (1921), 196–217.

[69] E. A. Carlen and E. H. Lieb, *A Minkowski type trace inequality and strong subadditivity of quantum entropy*, Differential Operators and Spectral Theory, pp. 59–68, Amer. Math. Soc. Transl. Ser. 2, 189, American Mathematical Society, Providence, RI, 1999.

[70] V. I. Chilin, P. G. Dodds, and F. A. Sukochev, *The Kadec-Klee property in symmetric spaces of measurable operators*, Israel J. Math. **97** (1997), 203–219.

[71] J. S. R. Chisholm, *Solution of linear integral equations using Padé approximants*, J. Math. Phys. **4** (1963), 1506–1510.

[72] S. Clark and F. Gesztesy, *Weyl-Titchmarsh M-function asymptotics, local uniqueness results, trace formulas, and Borg-type theorems for Dirac operators*, Trans. Amer. Math. Soc. **354** (2002), 3475–3534 [electronic].

[73] S. Clark, F. Gesztesy, H. Holden, and B. M. Levitan, *Borg-type theorems for matrix-valued Schrödinger operators*, J. Differential Equations **167** (2000), 181–210.

[74] J. A. Clarkson, *Uniformly convex spaces*, Trans. Amer. Math. Soc. **40** (1936), 396–414.

[75] S. Coleman, *There are no Goldstone bosons in two dimensions*, Comm. Math. Phys. **31** (1973), 259–264.

[76] J. M. Combes, *Time-dependent approach to nonrelativistic multichannel scattering*, Nuovo Cimento A (10) **64** (1969), 111–144.

[77] J. M. Combes, P. D. Hislop, and S. Nakamura, *The $L^p$-theory of the spectral shift function, the Wegner estimate, and the integrated density of states for some random operators*, Comm. Math. Phys. **218** (2001), 113–130.

[78] M. Combescure and J. Ginibre, *Scattering and local absorption for the Schrödinger operator*, J. Funct. Anal. **29** (1978), 54–73.

[79] A. Cooper and L. Rosen, *The weakly coupled Yukawa$_2$ field theory: Cluster expansion and Wightman axioms*, Trans. Amer. Math. Soc. **234** (1977), 1–88.

[80] M. Cwikel, *Weak type estimates for singular values and the number of bound states of Schrödinger operators*, Ann. of Math. (2) **106** (1977), 93–100.

[81] H. L. Cycon, R. G. Froese, W. Kirsch, and B. Simon, *Schrödinger Operators With Application to Quantum Mechanics and Global Geometry*, Springer-Verlag, Berlin, 1987.

[82] E. B. Davies and B. Simon, *Scattering theory for systems with different spatial asymptotics on the left and right*, Comm. Math. Phys. **63** (1978), 277–301.

[83] C. Davis, *Separation of two linear subspaces*, Acta Sci. Math. Szeged **19** (1958), 172–187.

[84] V. de Alfaro and T. Regge, *Potential Scattering*, North-Holland, Amsterdam, 1965.

[85] M. Dechamps-Gondim, F. Lust-Piquard, and H. Queffélec, *On the minorant properties in $C_p(H)$*, Pacific J. Math. **119** (1985), 89–101.

[86] P. A. Deift, *Classical Scattering Theory With a Trace Condition*, Ph.D. dissertation, Princeton, 1976.

[87] P. A. Deift, *Applications of a commutation formula*, Duke Math. J. **45** (1978), 267–310.

[88] P. A. Deift and B. Simon, *On the decoupling of finite singularities from the question of asymptotic completeness in two body quantum systems*, J. Funct. Anal. **23** (1976), 218–238.

[89] R. del Rio, *A forbidden set for embedded eigenvalues*, Proc. Amer. Math. Soc. **121** (1994), 77–82.

[90] R. del Rio, S. Jitomirskaya, Y. Last, and B. Simon, *Operators with singular continuous spectrum. IV. Hausdorff dimensions, rank one perturbations, and localization*, J. Anal. Math. **69** (1996), 153–200.

[91] R. del Rio, S. Jitomirskaya, N. Makarov, and B. Simon, *Singular continuous spectrum is generic*, Bull. Amer. Math. Soc. (N.S.) **31** (1994), 208–212.

[92] R. del Rio, N. Makarov, and B. Simon, *Operators with singular continuous spectrum. II. Rank one operators*, Comm. Math. Phys. **165** (1994), 59–67.

[93] R. del Rio, B. Simon, and G. Stolz, *Stability of spectral types for Sturm-Liouville operators*, Math. Res. Lett. **1** (1994), 437–450.

[94] S. J. Dilworth and Y.-P. Hsu, *On a property of Kadec-Klee type for quasi-normed unitary matrix spaces*, Far East J. Math. Sci. Special Volume, Part II (1996), 183–194.

[95] M. Dimassi and V. Petkov, *Spectral shift function and resonances for non-semi-bounded and Stark Hamiltonians*, J. Math. Pures Appl. (9) **82** (2003), 1303–1342.

[96] J. Dixmier, *Formes linéaires sur un anneau d'opérateurs*, Bull. Soc. Math. France **81** (1953), 9–39.

[97] P. G. Dodds, T. K. Dodds, and F. A. Sukochev, *Lifting of Kadec-Klee properties to symmetric spaces of measurable operators*, Proc. Amer. Math. Soc. **125** (1997), 1457–1467.

[98] P. G. Dodds and D. H. Fremlin, *Compact operators in Banach lattices*, Israel J. Math. **34** (1979), 287–320 (1980).

[99] J. Dombrowski, *Tridiagonal matrix representations of cyclic selfadjoint operators*, Pacific J. Math. **114** (1984), 325–334.

[100] J. Dombrowski, *Tridiagonal matrix representations of cyclic selfadjoint operators. II*, Pacific J. Math. **120** (1985), 47–53.

[101] J. Dombrowski, *Cyclic operators, commutators, and absolutely continuous measures*, Proc. Amer. Math. Soc. **100** (1987), 457–463.

[102] W. F. Donoghue, *On the perturbation of spectra*, Comm. Pure Appl. Math. **18** (1965), 559–579.

[103] W. F. Donoghue, *Monotone Matrix Functions and Analytic Continuation*, Springer, New York-Heidelberg-Berlin, 1974.

[104] M. Dostanić, *Trace formula for nonnuclear perturbations of selfadjoint operators*, Publ. Inst. Math. (Beograd) (N.S.) **54(68)** (1993), 71–79.

[105] N. Dunford and J. T. Schwartz, *Linear operators. Part II*, John Wiley, New York, 1988.

[106] H. Dym and H. P. McKean, *Fourier Series and Integrals*, Probability and Mathematical Statistics, No. 14, Academic Press, New York, 1972.

[107] M. S. P. Eastham, *The Spectral Theory of Periodic Differential Equations*, Scottish Academic Press, Edinburgh, 1973.

[108] E. G. Effros, *Why the circle is connected: An introduction to quantized topology*, Math. Intelligencer **11** (1989), 27–34.

[109] P. Enflo, *A counterexample to the approximation problem in Banach spaces*, Acta Math. **130** (1973), 309–317.

[110] V. Enss, *Asymptotic completeness for quantum mechanical potential scattering. I. Short range potentials*, Comm. Math. Phys. **61** (1978), 285–291.

[111] H. Epstein, *Remarks on two theorems of E. Lieb*, Comm. Math. Phys. **31** (1973), 317–325.

[112] K. Fan, *Maximum properties and inequalities for the eigenvalues of completely continuous operators*, Proc. Nat. Acad. Sci. U.S.A. **37** (1951), 760–766.

[113] C. Fernández and G. Raikov, *On the singularities of the magnetic spectral shift function at the Landau levels*, Ann. Henri Poincaré **5** (2004), 381–403.

[114] J. Fournier and B. Russo, *Abstract interpolation and operator-valued kernels*, J. London Math. Soc. (2) **16** (1977), 283–289.

[115] I. Fredholm, *Sur une classe d'equation fonctionelles*, Acta Math. **27** (1903), 365–390.

[116] R. Froese, *Asymptotic distribution of resonances in one dimension*, J. Differential Equations **137** (1997), 251–272.

[117] J. Fröhlich, F. Martinelli, E. Scoppola, and T. Spencer, *Constructive proof of localization in the Anderson tight binding model*, Comm. Math. Phys. **101** (1985), 21–46.

[118] J. Fröhlich, B. Simon, and T. Spencer, *Infrared bounds, phase transitions and continuous symmetry breaking*, Comm. Math. Phys. **50** (1976), 79–95.

[119] J. Fröhlich and T. Spencer, *Absence of diffusion in the Anderson tight binding model for large disorder or low energy*, Comm. Math. Phys. **88** (1983), 151–184.

[120] H. Furstenberg, *Noncommuting random products*, Trans. Amer. Math. Soc. **108** (1963), 377–428.

[121] C. R. Garibotti and M. Villani, *Padé approximant and the Jost function*, Nuovo Cimento A (10) **61** (1969), 747–765.

[122] R. Geisler, V. Kostrykin, and R. Schrader, *Concavity properties of Krein's spectral shift function*, Rev. Math. Phys. **7** (1995), 161–181.

[123] I. M. Gel'fand and B. M. Levitan, *On a simple identity for the characteristic values of a differential operator of the second order*, Dokl. Akad. Nauk SSSR (N.S.) **88** (1953), 593–596; English transl. in "Collected Papers, I" (edited by S. G. Gindikin, V. W. Guillemin, A. A. Kirillov, B. Kostant, and S. Sternberg), pp. 457–461, Springer-Verlag, Berlin, 1987,

[124] I. M. Gel'fand and N. Ya. Vilenkin, *Generalized Functions. Vol. 4: Applications of Harmonic Analysis*, Academic Press, New York, 1964.

[125] F. Gesztesy and K. A. Makarov, *The Ξ operator and its relation to Krein's spectral shift function*, J. Anal. Math. **81** (2000), 139–183.

[126] F. Gesztesy, K. A. Makarov, and A. K. Motovilov, *Monotonicity and concavity properties of the spectral shift function*, Stochastic Processes, Physics and Geometry: New Interplays, II (Leipzig, 1999), pp. 207–222, CMS Conf. Proc., 29, American Mathematical Society, Providence, RI, 2000.

[127] F. Gesztesy, K. A. Makarov, and S. N. Naboko, *The spectral shift operator*, Mathematical Results in Quantum Mechanics (Prague, 1998), pp. 59–90, Oper. Theory Adv. Appl., 108, Birkhäuser, Basel, 1999.

[128] F. Gesztesy, H. Holden, and B. Simon, *Absolute summability of the trace relation for certain Schrödinger operators*, Comm. Math. Phys. **168** (1995), 137–161.

[129] F. Gesztesy, H. Holden, B. Simon, and Z. Zhao, *Trace formulae and inverse spectral theory for Schrödinger operators*, Bull. Amer. Math. Soc. (N.S.) **29** (1993), 250–255.

[130] F. Gesztesy, H. Holden, B. Simon, and Z. Zhao, *Higher order trace relations for Schrödinger operators*, Rev. Math. Phys. **7** (1995), 893–922.

[131] F. Gesztesy and B. Simon, *Rank-one perturbations at infinite coupling*, J. Funct. Anal. **128** (1995), 245–252.

[132] F. Gesztesy and B. Simon, *The xi function*, Acta Math. **176** (1996), 49–71.

[133] F. Gesztesy and M. Zinchenko, *A Borg-type theorem associated with orthogonal polynomials on the unit circle*, preprint.

[134] I. C. Goh'berg and M. G. Kreĭn, *Introduction to the theory of linear nonselfadjoint operators*, Translations of Mathematical Monographs, Vol. 18, American Mathematical Society, Providence, RI, 1969.

[135] M. L. Goldberger and K. M. Watson, *Collision Theory*, Robert E. Krieger Publishing, Huntington, NY, 1975; corrected reprint of the 1964 edition.

[136] S. Golden, *Lower bounds for the Helmholtz function*, Phys. Rev. (2) **137** (1965), B1127–B1128.

[137] A. Ya. Gordon, *Exceptional values of the boundary phase for the Schrödinger equation on the semi-axis*, Russian Math. Surveys **47** (1992), 260–261.

[138] A. Ya. Gordon, *Pure point spectrum under 1-parameter perturbations and instability of Anderson localization*, Comm. Math. Phys. **164** (1994), 489–505.

[139] L. Gross, *Existence and uniqueness of physical ground states*, J. Funct. Anal. **10** (1972), 52–109.

[140] L. Gross, *On the formula of Matthews and Salam*, J. Funct. Anal. **25** (1977), 162–209.

[141] A. Grossmann and T. T. Wu, *Schrödinger scattering amplitude. I*, J. Math. Phys. **2** (1961), 710–713.

[142] A. Grossmann and T. T. Wu, *Schrödinger scattering amplitude. III*, J. Math. Phys. **3** (1962), 684–689.

[143] A. Grothendieck, *Produits tensoriels topologiques et espaces nucléaires*, Mem. Amer. Math. Soc. **1955** (1955), no. 16, 140.

[144] A. Grothendieck, *La théorie de Fredholm*, Bull. Soc. Math. France **84** (1956), 319–384.

[145] H. R. Grümm, *Two theorems about* $\mathcal{C}_p$, Rep. Math. Phys. **4** (1973), 211–215.

[146] O. Hanner, *On the uniform convexity of* $L^p$ *and* $l^p$, Ark. Mat. **3** (1956), 239–244.

[147] G. H. Hardy, J. E. Littlewood, and G. Pólya, *Inequalities*, 2nd edition, Cambridge University Press, Cambridge, 1952.

[148] F. Hiai, *Concavity of certain matrix trace functions*, Taiwanese J. Math. **5** (2001), 535–554.

[149] D. Hilbert, *Grundzüge einer allgemeinen Theorie der linearen Integralgleichungen. I*, Nach. Wiss. Math. Phys. Gott. (1904), 49–91.

[150] A. Horn, *On the singular values of a product of completely continuous operators*, Proc. Nat. Acad. Sci. U.S.A. **36** (1950), 374–375.

[151] D. Hundertmark, R. Killip, S. Nakamura, P. Stollmann, and I. Veselic, *Bounds on the spectral shift function and the density of states*, preprint.

[152] D. Hundertmark and B. Simon, *An optimal* $L^p$*-bound on the Krein spectral shift function*, Dedicated to the memory of Thomas H. Wolff, J. Anal. Math. **87** (2002), 199–208.

[153] T. Ikebe, *Eigenfunction expansions associated with the Schroedinger operators and their applications to scattering theory*, Arch. Rational Mech. Anal. **5** (1960), 1–34 (1960).

[154] V. Ya. Ivrii and M. A. Shubin, *On the asymptotic behavior of the spectral shift function*, Soviet Math. Dokl. **25** (1982), 332–334; Russian original in Dokl. Akad. Nauk SSSR **263** (1982), 283–284.

[155] V. Jakšić, *Topics in spectral theory*, Open Quantum Systems (Grenoble, 2003), preprint.

[156] V. Jakšić and Y. Last, *Spectral structure of Anderson type Hamiltonians*, Invent. Math. **141** (2000), 561–577.

[157] V. Jakšić and Y. Last, *Surface states and spectra*, Comm. Math. Phys. **218** (2001), 459–477.

[158] V. Jakšić and Y. Last, *A new proof of Poltoratskii's theorem*, J. Funct. Anal. **215** (2004), 103–110.

[159] V. Jakšić and Y. Last, *Simplicity of singular spectrum in Anderson type Hamiltonians*, submitted.

[160] V. Jakšić and S. Molchanov, *Localization of surface spectra*, Comm. Math. Phys. **208** (1999), 153–172.

[161] V. A. Javrjan, *A certain inverse problem for Sturm-Liouville operators*, Izv. Akad. Nauk Armjan. SSR Ser. Mat. **6** (1971), 246–251 [Russian].

[162] S. Jitomirskaya and B. Simon, *Operators with singular continuous spectrum. III. Almost periodic Schrödinger operators*, Comm. Math. Phys. **165** (1994), 201–205.

[163] W. B. Johnson, H. König, B. Maurey, and J. R. Retherford, *Eigenvalues of p-summing and* $l_p$*-type operators in Banach spaces*, J. Funct. Anal. **32** (1979), 353–380.

[164] T. Kato, *Perturbation Theory for Linear Operators*, Grundlehren der Mathematischen Wissenschaften, Band 132, Springer-Verlag, New York, 1966; 2nd edition, 1976.

[165] T. Kato, *Trotter's product formula for an arbitrary pair of self-adjoint contraction semigroups*, Topics in Functional Analysis, Adv. in Math. Suppl. Stud., Vol. 3, pp. 185–195, Academic Press, New York, 1978.

[166] T. Kato, unpublished.

[167] T. Kato and K. Masuda, *Trotter's product formula for nonlinear semigroups generated by the subdifferentials of convex functionals*, J. Math. Soc. Japan **30** (1978), 169–178.

[168] Y. Katznelson, *An Introduction to Harmonic Analysis*, corrected edition, Dover, New York, 1976.

[169] M. Klaus, *On the bound state of Schrödinger operators in one dimension*, Ann. Physics **108** (1977), 288–300.

[170] M. Klaus, private communication.

[171] A. Klein and B. Russo, *Sharp inequalities for Weyl operators and Heisenberg groups*, Math. Ann. **235** (1978), 175–194.

[172] L. S. Koplienko, *The trace formula for perturbations of nonnuclear type*, Sibirsk. Mat. Zh. **25** (1984), 62–71 [Russian].

[173] V. Kostrykin, *Concavity of eigenvalue sums and the spectral shift function*, J. Funct. Anal. **176** (2000), 100–114.

[174] V. Kostrykin and R. Schrader, *Scattering theory approach to random Schrödinger operators in one dimension*, Rev. Math. Phys. **11** (1999), 187–242.

[175] V. Kostrykin and R. Schrader, *The density of states and the spectral shift density of random Schrödinger operators*, Rev. Math. Phys. **12** (2000), 807–847.

[176] S. Kotani, *Ljapunov indices determine absolutely continuous spectra of stationary random one-dimensional Schrödinger operators*, Stochastic Analysis (Katata/Kyoto, 1982), vol. 32, pp. 225–247, North-Holland, Amsterdam, 1984.

[177] G. Köthe, *Topological Vector Spaces. I*, Grundlehren der Mathematischen Wissenschaften, Band 159, Springer-Verlag, New York, 1969.

[178] M. G. Kreĭn, *Perturbation determinants and a formula for the traces of unitary and self-adjoint operators*, Soviet Math. Dokl. **3** (1962), 707–710; Russin original in Dokl. Akad. Nauk SSSR **144** (1962), 68–271.

[179] M. G. Kreĭn and V. A. Yavryan, *Spectral shift functions that arise in perturbations of a positive operator*, J. Operator Theory **6** (1981), 155–191 [Russian].

[180] R. A. Kunze, *$L_p$ Fourier transforms on locally compact unimodular groups*, Trans. Amer. Math. Soc. **89** (1958), 519–540.

[181] S. T. Kuroda, *On a theorem of Weyl-von Neumann*, Proc. Japan Acad. **34** (1958), 11–15.

[182] S. T. Kuroda, *On a paper of Green and Lanford*, J. Math. Phys. **3** (1962), 933–935.

[183] S. T. Kuroda, *Scattering theory for differential operators. I. Operator theory*, J. Math. Soc. Japan **25** (1973), 75–104.

[184] S. T. Kuroda, *Scattering theory for differential operators. II. Self-adjoint elliptic operators*, J. Math. Soc. Japan **25** (1973), 222–234.

[185] S. Kwapień, *On a theorem of L. Schwartz and its applications to absolutely summing operators*, Studia Math. **38** (1970), 193–201.

[186] T. Lalesco, *Une théorème dur les noyaux composes*, Bull. Acad. Sci. Roumaine **3** (1915), 271–272.

[187] H. Langer, H. S. V. de Snoo, and V. A. Yavryan, *A relation for the spectral shift function of two self-adjoint extensions*, Recent Advances in Operator Theory and Related Topics (Szeged, 1999), pp. 437–445, Oper. Theory Adv. Appl., 127, Birkhäuser, Basel, 2001.

[188] Y. Last, *A relation between a.c. spectrum of ergodic Jacobi matrices and the spectra of periodic approximants*, Comm. Math. Phys. **151** (1993), 183–192.

[189] P. D. Lax and R. S. Phillips, *Scattering Theory*, Pure and Applied Mathematics, Vol. 26, Academic Press, New York, 1967.

[190] A. Lenard, *Generalization of the Golden-Thompson inequality* $\mathrm{Tr}(e^A e^B) \geq \mathrm{Tr}\, e^{A+B}$, Indiana Univ. Math. J. **21** (1971/1972), 457–467.

[191] B. M. Levitan and M. G. Gasymov, *Determination of a differential equation by two spectra*, Russian Math. Surveys **19** (1964), 1–63; Russian original in Uspekhi Mat. Nauk **19** (1964) no. 2 (116), 3–63.

[192] T. Leżański, *The Fredholm theory of linear equations in Banach spaces*, Studia Math. **13** (1953), 244–276.

[193] V. B. Lidskii, *On the characteristic numbers of the sum and product of symmetric matrices*, Dokl. Akad. Nauk SSSR (N.S.) **75** (1950), 769–772 [Russian].

[194] V. B. Lidskii, *Non-selfadjoint operators with a trace*, Dokl. Akad. Nauk SSSR **125** (1959), 485–487 [Russian].

[195] E. H. Lieb, *The classical limit of quantum spin systems*, Comm. Math. Phys. **31** (1973), 327–340.

[196] E. H. Lieb, *Convex trace functions and the Wigner-Yanase-Dyson conjecture*, Advances in Math. **11** (1973), 267–288.

[197] E. H. Lieb, *Some convexity and subadditivity properties of entropy*, Bull. Amer. Math. Soc. **81** (1975), 1–13.

[198] E. H. Lieb, *Bounds on the eigenvalues of the Laplace and Schroedinger operators*, Bull. Amer. Math. Soc. **82** (1976), 751–753.

[199] E. H. Lieb, *Inequalities for some operator and matrix functions*, Advances in Math. **20** (1976), 174–178.

[200] E. H. Lieb and M. B. Ruskai, *Proof of the strong subadditivity of quantum-mechanical entropy* (with an appendix by B. Simon), J. Math. Phys. **14** (1973), 1938–1941.

[201] E. H. Lieb, B. Simon, and A. S. Wightman (editors), *Studies in Mathematical Physics: Essays in Honor of Valentine Bargmann*, Princeton University Press, Princeton, NJ, 1971.

[202] E. H. Lieb and W. Thirring, *Bound for the kinetic energy of fermions which proves the stability of matter*, Phys. Rev. Lett. **35** (1975), 687–689.

[203] E. H. Lieb and W. Thirring, *Inequalities for the moments of the eigenvalues of the Schrödinger Hamiltonian and their relation to Sobolev inequalities*, Studies in Mathematical Physics: Essays in Honor of Valentine Bargmann (edited by E. H. Lieb, B. Simon, and A. S. Wightman), pp. 269–303, Princeton University Press, Princeton, NJ, 1971.

[204] J. Lindenstrauss and L. Tzafriri, *Classical Banach Spaces. I*, Sequence Spaces, Ergebnisse der Mathematik undihrer Grenzgebiete, Vol. 92, Springer-Verlag, Berlin, 1977.

[205] J.-L. Lions, *Théorèmes de trace et d'interpolation. I*, Ann. Scuola Norm. Sup. Pisa (3) **13** (1959), 389–403.

[206] J.-L. Lions, *Théorèmes de trace et d'interpolation. II*, Ann. Scuola Norm. Sup. Pisa (3) **14** (1960), 317–331.

[207] B. A. Lippmann and J. Schwinger, *Variational principles for scattering processes. I*, Phys. Rev. **79** (1950), 469–480.

[208] R. A. H. Lorentz and P. A. Rejto, *Some integral operators of trace class*, Acta Sci. Math. (Szeged) **36** (1974), 91–105.

[209] W. V. Lovitt, *Linear Integral Equations*, Dover, New York, 1950.

[210] K. Löwner, *Über monotone Matrixfunktionen*, Math. Z. **38** (1934), 177–216.

[211] J. Magnen and R. Sénéor, *The Wightman axioms for the weakly coupled Yukawa model in two dimensions*, Comm. Math. Phys. **51** (1976), 297–313.

[212] V. A. Marčenko, *Some questions of the theory of one-dimensional linear differential operators of the second order. I*, Amer. Math. Soc. Trans. Ser. 2 **101**, (1973), 1–104; Russian original in Trudy Moskov. Mat. Obšč. **1** (1952), 327–420.

[213] A. S. Markus, *Eigenvalues and singular values of the sum and product of linear operators*, Soviet Math. Dokl. **3** (1962), 1238–1241; Russian original in Dokl. Akad. Nauk SSSR **146** (1962), 34–36.

[214] A. S. Markus and V. I. Macaev, *Analogues of the Weyl inequalities, and trace theorems in a Banach space*, Math. USSR Sb. **15** (1971), 299–312; Russian original in Mat. Sb. (N.S.) **86(128)** (1971), 299–313.

[215] A. W. Marshall and I. Olkin, *Inequalities: Theory of Majorization and Its Applications*, Academic Press, New York-London, 1979.

[216] A. Martin, *Bound states in the strong coupling limit*, Helv. Phys. Acta **45** (1972), 140–148.

[217] P. T. Matthews and A. Salam, *The Green's function of quantised field*, Nuovo Cimento (9) **12** (1954), 563–565.

[218] O. A. McBryan, *Volume dependence of Schwinger function in the Yukawa*$_2$ *quantum field theory*, Comm. Math. Phys. **45** (1975), 279–294.

[219] C. A. McCarthy, $c_p$, Israel J. Math. **5** (1967), 249–271.

[220] N. D. Mermin and H. Wagner, *Absence of ferromagnetism or antiferromagnetism in one- or two-dimensional isotropic Heisenberg models*, Phys. Rev. Lett. **17** (1966), 1133–1136.

[221] R. Merris and J. A. Dias da Silva, *Generalized Schur functions*, J. Algebra **35** (1975), 442–448.

[222] N. Minami, *Local fluctuation of the spectrum of a multidimensional Anderson tight binding model*, Comm. Math. Phys. **177** (1996), 709–725.

[223] B. S. Mitjagin, *Normed ideals of intermediate type*, Amer. Math. Soc. Transl. **63** (1967), 180–194; Russian original in Izv. Akad. Nauk SSSR Ser. Mat. **28** (1964), 819–832.

[224] A. Mohapatra and K. B. Sinha, *Spectral shift function and trace formula for unitaries—a new proof*, Integral Equations Operator Theory **24** (1996), 285–297.

[225] J. D. Morgan, *Schrödinger operators whose potentials have separated singularities*, J. Operator Theory **1** (1979), 109–115.

[226] S. Nakamura, *Spectral shift function for trapping energies in the semiclassical limit*, Comm. Math. Phys. **208** (1999), 173–193.

[227] H. Neidhardt, *Spectral shift function and Hilbert-Schmidt perturbation: Extensions of some work of L. S. Koplienko*, Math. Nachr. **138** (1988), 7–25.

[228] E. Nelson, *Notes on non-commutative integration*, J. Funct. Anal. **15** (1974), 103–116.

[229] R. G. Newton, *Scattering theory of waves and particles*, McGraw-Hill, New York, 1966.

[230] V. I. Oseledec, *A multiplicative ergodic theorem. Characteristic Ljapunov, exponents of dynamical systems*, Trans. Moscow Math. Soc. **19** (1968), 197–231; Russian original in Trudy Moskov. Mat. Obšč. **19** (1968), 179–210.

[231] K. Osterwalder and R. Schrader, *Euclidean Fermi fields and a Feynman-Kac formula for boson-fermion models*, Helv. Phys. Acta **46** (1973/74), 277–302.

[232] D. B. Pearson, *A generalization of the Birman trace theorem*, J. Funct. Anal. **28** (1978), 182–186.

[233] J. Peetre, preprint, 1978.

[234] V. V. Peller, *Hankel operators of class* $\mathfrak{S}_p$ *and their applications* (*rational approximation, Gaussian processes, the problem of majorization of operators*), Mat. Sb. (N.S.) **113(155)** (1980), 538–581, 637 [Russian].

[235] V. V. Peller, *Hankel Operators and Their Applications*, Springer Monographs in Mathematics, Springer-Verlag, New York, 2003.

[236] A. Pietsch, *Einige neue Klassen von kompakten linearen Abbildungen*, Rev. Math. Pures Appl. (Bucarest) **8** (1963), 427–447.

[237] A. Pietsch, *Absolut p-summierende Abbildungen in normierten Räumen*, Studia Math. **28** (1966/1967), 333–353.

[238] G. Pisier, *The Volume of Convex Bodies and Banach Space Geometry*, Cambridge Tracts in Mathematics, 94, Cambridge University Press, Cambridge, 1989.

[239] L. D. Pitt, *A compactness condition for linear operators of function spaces*, J. Operator Theory **1** (1979), 49–54.

[240] J. Plemlj, *Zur Theorie der Fredholmschen Funktionalgleichung*, Monatsh. Math. Phys. **15** (1904), 93–128.

[241] H. Poincaré, *Remarques diverses sur l'equation de Fredholm*, Acta Math. **33** (1910), 57–86.

[242] A. G. Poltoratskii, *Boundary behavior of pseudocontinuable functions*, St. Petersburg Math. J. **5** (1994), 389–406; Russian original in Algebra i Analiz **5** (1993), 189–210.

[243] A. G. Poltoratskii, *Krein's spectral shift and perturbations of spectra of rank one*, St. Petersburg Math. J. **10** (1999), 833–859; Russian original in Algebra i Analiz **10** (1998), 143–183.

[244] A. B. Pushnitskii, *A representation for the spectral shift function in the case of perturbations of fixed sign*, St. Petersburg Math. J. **9** (1998), 1181–1194; Russian original in Algebra i Analiz **9** (1997), 197–213.

[245] A. B. Pushnitskii, *Integral estimates for the spectral shift function*, St. Petersburg Math. J. **10** (1999), 1047–1070; Russian original in Algebra i Analiz **10** (1998), 198–233.

[246] A. B. Pushnitskii, *Spectral shift function of the Schrödinger operator in the large coupling constant limit*, Comm. Partial Differential Equations **25** (2000), 703–736.

[247] A. B. Pushnitskii, *The spectral shift function and the invariance principle*, J. Funct. Anal. **183** (2001), 269–320.

[248] A. B. Pushnitskii and M. V. Ruzhanskii, *The spectral shift function of the Schrödinger operator in the large coupling constant limit*, Funct. Anal. Appl. **36** (2002), 250–252; Russian original in Funktsional. Anal. i Prilozhen. **36** (2002), 93–95.

[249] A. B. Pushnitskii and M. V. Ruzhanskii, *Spectral shift function of the Schrödinger operator in the large coupling constant limit. II. Positive perturbations*, Comm. Partial Differential Equations **27** (2002), 1373–1405.

[250] M. Reed and B. Simon, *Methods of Modern Mathematical Physics. I. Functional Analysis*, Academic Press, New York, 1972.

[251] M. Reed and B. Simon, *Methods of Modern Mathematical Physics. II. Fourier Analysis, Self-adjointness*, Academic Press, New York, 1975.

[252] M. Reed and B. Simon, *The scattering of classical waves from inhomogeneous media*, Math. Z. **155** (1977), 163–180.

[253] M. Reed and B. Simon, *Methods of Modern Mathematical Physics. III. Scattering Theory*, Academic Press, New York, 1979.

[254] M. Reed and B. Simon, *Methods of Modern Mathematical Physics. IV. Analysis of Operators*, Academic Press, New York, 1978.

[255] J. R. Retherford, *Applications of Banach ideals of operators*, Bull. Amer. Math. Soc. **81** (1975), 978–1012.

[256] J. Ringrose, *Fredholm Theory*, Van Nostrand, 1974.

[257] D. Robert, *Semiclassical asymptotics for the spectral shift function*, Differential Operators and Spectral Theory, pp. 187–203, Amer. Math. Soc. Transl. Ser. 2, 189, American Mathematical Society, Providence, RI, 1999.

[258] L. Rosen, *Positive powers of positive positive-definite matrices*, Canad. J. Math. **48** (1996), 196–209.

[259] C. V. Rosenbljum, *The distribution of the discrete spectrum for singular differential operators*, Soviet Math. Dokl. **13** (1972), 245–249; Russian original in Dokl. Akad. Nauk SSSR **202** (1972), 1012–1015.

[260] S. Yu. Rotfel'd, *The singular numbers of the sum of completely continuous operators*, Topics in Mathematical Physics, Vol. 3: Spectral Theory, pp. 73–78, Consultants Bureau, New York-London, 1969.

[261] W. Rudin, *Real and Complex Analysis*, 3rd edition, McGraw-Hill, New York, 1987.

[262] D. Ruelle, *Statistical Mechanics: Rigorous Results*, W. A. Benjamin, New York-Amsterdam, 1969.

[263] D. Ruelle, *Ergodic theory of differentiable dynamical systems*, Inst. Hautes Études Sci. Publ. Math. (1979), no. 50, 27–58.

[264] M. B. Ruskai, *Lieb's simple proof of concavity of* $(A, B) \mapsto \operatorname{Tr} A^p K^\dagger B^{1-p} K$ *and remarks on related inequalities*, preprint.

[265] B. Russo, *The norm of the* $L^p$*-Fourier transform on unimodular groups*, Trans. Amer. Math. Soc. **192** (1974), 293–305.

[266] B. Russo, *The norm of the* $L^p$*-Fourier transform. II*, Canad. J. Math. **28** (1976), 1121–1131.

[267] B. Russo, *On the Hausdorff-Young theorem for integral operators*, Pacific J. Math. **68** (1977), 241–253.

[268] B. Russo, *The norm of the* $L^p$*-Fourier transform. III. Compact extensions*, J. Funct. Anal. **30** (1978), 162–178.

[269] A. F. Ruston, *Direct products of Banach spaces and linear functional equations*, Proc. London Math. Soc. (3) **1** (1951), 327–384.

[270] A. F. Ruston, *On the Fredholm theory of integral equations for operators belonging to the trace class of a general Banach space*, Proc. London Math. Soc. (2) **53** (1951), 109–124.

[271] A. V. Rybkin, *The spectral shift function, the characteristic function of a contraction and a generalized integral*, Russian Acad. Sci. Sb. Math. **83** (1995), 237–281; Russian original in Mat. Sb. **185** (1994), 91–144.

[272] A. V. Rybkin, *On A-integrability of the spectral shift function of unitary operators arising in the Lax-Phillips scattering theory*, Duke Math. J. **83** (1996), 683–699.

[273] V. A. Sadovnichii and V. E. Podol'skii, *Traces of operators with relative trace-class perturbation*, Dokl. Akad. Nauk **378** (2001), 324–325 [Russian].

[274] V. A. Sadovnichii and V. E. Podol'skii, *Traces of operators with a relatively compact perturbation*, Sb. Math. **193** (2002), 279–302; Russian original in Mat. Sb. **193** (2002), 129–152.

[275] O. Safronov, *Spectral shift function in the large coupling constant limit*, J. Funct. Anal. **182** (2001), 151–169.

[276] S. Saks, *Theory of the Integral*, Hafner, New York, 1937.

[277] M. Scadron, S. Weinberg, and J. Wright, *Functional analysis and scattering theory*, Phys. Rev. (2) **135** (1964), B202–B207.

[278] R. Schatten, *Norm Ideals of Completely Continuous Operators*, Ergebnisse der Mathematik und ihrer Grenzgebiete. N. F., Heft 27, Springer-Verlag, Berlin, 1960.

[279] R. Schatten and J. von Neumann, *The cross-space of linear transformations. II*, Ann. of Math. (2) **47** (1946), 608–630.

[280] R. Schatten and J. von Neumann, *The cross-space of linear transformations. III*, Ann. of Math. (2) **49** (1948), 557–582.

[281] I. Schur, *Über die charakteristischen Wurzeln einer linearen Substitution mit einer Anwendung auf die Theorie der Integralgleichung*, Math. Ann. **66** (1909), 488–510.

[282] J. Schwinger, *On the bound states of a given potential*, Proc. Nat. Acad. Sci. U.S.A. **47** (1961), 122–129.

[283] I. E. Segal, *A non-commutative extension of abstract integration*, Ann. of Math. (2) **57** (1953), 401–457.

[284] I. E. Segal, *Notes towards the construction of nonlinear relativistic quantum fields. III. Properties of the $C^*$-dynamics for a certain class of interactions*, Bull. Amer. Math. Soc. **75** (1969), 1390–1395.

[285] E. Seiler, *Schwinger functions for the Yukawa model in two dimensions with space-time cutoff*, Comm. Math. Phys. **42** (1975), 163–182.

[286] E. Seiler and B. Simon, *Bounds in the Yukawa$_2$ quantum field theory: Upper bound on the pressure, Hamiltonian bound and linear lower bound*, Comm. Math. Phys. **45** (1975), 99–114.

[287] E. Seiler and B. Simon, *An inequality among determinants*, Proc. Nat. Acad. Sci. U.S.A. **72** (1975), 3277–3278.

[288] E. Seiler and B. Simon, *On finite mass renormalizations in the two-dimensional Yukawa model*, J. Math. Phys. **16** (1975), 2289–2293.

[289] E. Seiler and B. Simon, *Nelson's symmetry and all that in the Yukawa2 and $(\phi^4)_3$ field theories*, Ann. Physics **97** (1976), 470–518.

[290] T. Shirai, *A trace formula for discrete Schrödinger operators*, Publ. Res. Inst. Math. Sci. **34** (1998), 27–41.

[291] B. Simon, *Convergence of regularized, renormalized perturbation series for super-renormalizable field theories*, Nuovo Cimento A (10) **59** (1969), 199–214.

[292] B. Simon, *Quantum Mechanics for Hamiltonians Defined as Quadratic Forms*, Princeton University Press, Princeton, N.J., 1971.

[293] B. Simon, *Convergence theorems for entropy*, appendix to [**200**], J. Math. Phys. **14** (1973), 1938–1941.

[294] B. Simon, *On the number of bound states of two-body Schrödinger operators: A review*, Constructive Quantum Field Theory (Proc. "Ettore Majorana" International School of Mathematical Physics, Erice, Sicily, 1973; edited by G. Velo and A. Wightman), pp. 305–326, Lecture Notes in Physics, Vol. 25, Springer-Verlag, Berlin-New York, 1973.

[295] B. Simon, *The $P(\phi)_2$ Euclidean (Quantum) Field Theory*, Princeton University Press, Princeton, N.J., 1974.

[296] B. Simon, *Analysis with weak trace ideals and the number of bound states of Schrödinger operators*, Trans. Amer. Math. Soc. **224** (1976), 367–380.

[297] B. Simon, *Analysis with weak trace ideals and the number of bound states of Schrödinger operators*, Trans. Amer. Math. Soc. **224** (1976), 367–380.

[298] B. Simon, *The bound state of weakly coupled Schrödinger operators in one and two dimensions*, Ann. Physics **97** (1976), 279–288.

[299] B. Simon, *Geometric methods in multiparticle quantum systems*, Comm. Math. Phys. **55** (1977), 259–274.

[300] B. Simon, *Notes on infinite determinants of Hilbert space operators*, Advances in Math. **24** (1977), 244–273.

[301] B. Simon, *A canonical decomposition for quadratic forms with applications to monotone convergence theorems*, J. Funct. Anal. **28** (1978), 377–385.

[302] B. Simon, *N-body scattering in the two-cluster region*, Comm. Math. Phys. **58** (1978), 205–210.

[303] B. Simon, *Functional Integration and Quantum Physics*, Pure and Applied Mathematics, Vol. 86, Academic Press, New York, 1979; 2nd edition, AMS Chelsea Publishing, American Mathematical Society, Providence, RI, 2005.

[304] B. Simon, *Convergence in trace ideals*, Proc. Amer. Math. Soc. **83** (1981), 39–43.

[305] B. Simon, *Kotani theory for one-dimensional stochastic Jacobi matrices*, Comm. Math. Phys. **89** (1983), 227–234.

[306] B. Simon, *Localization in general one-dimensional random systems. I. Jacobi matrices*, Comm. Math. Phys. **102** (1985), 327–336.

[307] B. Simon, *Cyclic vectors in the Anderson model*, Rev. Math. Phys. **6** (1994), 1183–1185.

[308] B. Simon, *Operators with singular continuous spectrum. I. General operators*, Ann. of Math. (2) **141** (1995), 131–145.

[309] B. Simon, *Spectral averaging and the Krein spectral shift*, Proc. Amer. Math. Soc. **126** (1998), 1409–1413.

[310] B. Simon, *Resonances in one dimension and Fredholm determinants*, J. Funct. Anal. **178** (2000), 396–420.

[311] B. Simon, *Orthogonal Polynomials on the Unit Circle, Part 1: Classical Theory*, AMS Colloquium Series, American Mathematical Society, Providence, RI, 2005.

[312] B. Simon, *Aizenman's theorem for orthogonal polynomials on the unit circle*, to appear in Const. Approx.

[313] B. Simon and T. Spencer, *Trace class perturbations and the absence of absolutely continuous spectra*, Comm. Math. Phys. **125** (1989), 113–125.

[314] B. Simon and T. Wolff, *Singular continuous spectrum under rank one perturbations and localization for random Hamiltonians*, Comm. Pure Appl. Math. **39** (1986), 75–90.

[315] K. B. Sinha and A. N. Mohapatra, *Spectral shift function and trace formula*, Spectral and Inverse Spectral Theory (Bangalore, 1993), Proc. Indian Acad. Sci. Math. Sci. **104** (1994), 819–853.

[316] F. Smithies, *The Fredholm theory of integral equations*, Duke Math. J. **8** (1941), 107–130.

[317] F. Smithies, *Integral Equations*, Cambridge Tracts in Mathematics and Mathematical Physics, no. 49, Cambridge University Press, New York, 1958.

[318] A. V. Sobolev, *Efficient bounds for the spectral shift function*, Ann. Inst. H. Poincaré Phys. Théor. **58** (1993), 55–83.

[319] E. M. Stein and G. Weiss, *Introduction to Fourier Analysis on Euclidean Spaces*, Princeton University Press, Princeton, NJ, 1971.

[320] W. F. Stinespring, *A sufficient condition for an integral operator to have a trace*, J. Reine Angew. Math. **200** (1958), 200–207.

[321] M. Stoiciu, *The statistical distribution of the zeros of random paraorthogonal polynomials on the unit circle*, preprint.

[322] S. Strătilă, *Modular Theory in Operator Algebras*, Abacus Press, Tunbridge Wells, 1981.

[323] R. S. Strichartz, *Multipliers on fractional Sobolev spaces*, J. Math. Mech. **16** (1967), 1031–1060.

[324] K. Symanzik, *Proof and refinements of an inequality of Feynman*, J. Math. Phys. **6** (1965), 1155–1156.

[325] M. Takesaki, *Theory of Operator Algebras. I*, Springer-Verlag, New York-Heidelberg, 1979.

[326] H. Tamura, *The asymptotic eigenvalue distribution for non-smooth elliptic operators*, Proc. Japan Acad. **50** (1974), 19–22.

[327] C. J. Thompson, *Inequality with applications in statistical mechanics*, J. Math. Phys. **6** (1965), 1812–1813.

[328] C. J. Thompson, *Inequalities and partial orders on matrix spaces*, Indiana Univ. Math. J. **21** (1971/72), 469–480.

[329] E. C. Titchmarsh, *The Theory of Functions*, Oxford University Press, Oxford, 1932.

[330] E. C. Titchmarsh, *Eigenfunction Expansions Associated With Second-Order Differential Equations. Part I*, 2nd edition, Clarendon Press, Oxford, 1962.

[331] N. Tomczak-Jaegermann, *The moduli of smoothness and convexity and the Rademacher averages of trace classes* $S_p(1 \leq p < \infty)$, Studia Math. **50** (1974), 163–182.

[332] H. F. Trotter, *On the product of semi-groups of operators*, Proc. Amer. Math. Soc. **10** (1959), 545–551.

[333] A. Uhlmann, *Relative entropy and the Wigner-Yanase-Dyson-Lieb concavity in an interpolation theory*, Comm. Math. Phys. **54** (1977), 21–32.

[334] A. Uhlmann, unpublished.

[335] A. Vasy and X. P. Wang, *Smoothness and high energy asymptotics of the spectral shift function in many-body scattering*, Comm. Partial Differential Equations **27** (2002), 2139–2186.

[336] H. von Dreifus and A. Klein, *A new proof of localization in the Anderson tight binding model*, Comm. Math. Phys. **124** (1989), 285–299.

[337] F. Wegner, *Bounds on the density of states in disordered systems*, Z. Phys. B **44** (1981), 9–15.

[338] J. Weidmann, *Integraloperatoren der Spurklasse*, Math. Ann. **163** (1966), 340–345.

[339] H. Weyl, *Über beschränkte quadratische Formen, deren Differenz vollstetig ist*, Rend. Circ. Math. Palermo **27** (1909), 373–392.

[340] H. Weyl, *Inequalities between the two kinds of eigenvalues of a linear transformation*, Proc. Nat. Acad. Sci. U.S.A. **35** (1949), 408–411.

[341] E. P. Wigner and M. M. Yanase, *Information contents of distributions*, Proc. Nat. Acad. Sci. U.S.A. **49** (1963), 910–918.

[342] E. P. Wigner and M. M. Yanase, *On the positive semidefinite nature of a certain matrix expression*, Canad. J. Math. **16** (1964), 397–406.

[343] V. A. Yavryan, *On M. G. Krein's spectral shift function for canonical systems of differential equations*, Differential Operators and Related Topics, Vol. I (Odessa, 1997), pp. 393–417, Oper. Theory Adv. Appl., 117, Birkhäuser, Basel, 2000.

[344] C. Zemach and A. Klein, *The Born expansion in nonrelativistic quantum theory*, Nuovo Cimento (10) **10** (1958), 1078–1087.

# Index

# Titles in This Series

120 **Barry Simon,** Trace Ideals and Their Applications, 2005

119 **Tian Ma and Shouhong Wang,** Geometric theory of incompressible flows with applications to fluid dynamics, 2005

118 **Alexandru Buium,** Arithmetic differential equations, 2005

117 **Volodymyr Nekrashevych,** Self-similar groups, 2005

116 **Alexander Koldobsky,** Fourier analysis in convex geometry, 2005

115 **Carlos Julio Moreno,** Advanced analytic number theory: L-functions, 2005

114 **Gregory F. Lawler,** Conformally invariant processes in the plane, 2005

113 **William G. Dwyer, Philip S. Hirschhorn, Daniel M. Kan, and Jeffrey H. Smith,** Homotopy limit functors on model categories and homotopical categories, 2004

112 **Michael Aschbacher and Stephen D. Smith,** The classification of quasithin groups II. Main theorems: The classification of simple QTKE-groups, 2004

111 **Michael Aschbacher and Stephen D. Smith,** The classification of quasithin groups I. Structure of strongly quasithin $K$-groups, 2004

110 **Bennett Chow and Dan Knopf,** The Ricci flow: An introduction, 2004

109 **Goro Shimura,** Arithmetic and analytic theories of quadratic forms and Clifford groups, 2004

108 **Michael Farber,** Topology of closed one-forms, 2004

107 **Jens Carsten Jantzen,** Representations of algebraic groups, 2003

106 **Hiroyuki Yoshida,** Absolute CM-periods, 2003

105 **Charalambos D. Aliprantis and Owen Burkinshaw,** Locally solid Riesz spaces with applications to economics, second edition, 2003

104 **Graham Everest, Alf van der Poorten, Igor Shparlinski, and Thomas Ward,** Recurrence sequences, 2003

103 **Octav Cornea, Gregory Lupton, John Oprea, and Daniel Tanré,** Lusternik-Schnirelmann category, 2003

102 **Linda Rass and John Radcliffe,** Spatial deterministic epidemics, 2003

101 **Eli Glasner,** Ergodic theory via joinings, 2003

100 **Peter Duren and Alexander Schuster,** Bergman spaces, 2004

99 **Philip S. Hirschhorn,** Model categories and their localizations, 2003

98 **Victor Guillemin, Viktor Ginzburg, and Yael Karshon,** Moment maps, cobordisms, and Hamiltonian group actions, 2002

97 **V. A. Vassiliev,** Applied Picard-Lefschetz theory, 2002

96 **Martin Markl, Steve Shnider, and Jim Stasheff,** Operads in algebra, topology and physics, 2002

95 **Seiichi Kamada,** Braid and knot theory in dimension four, 2002

94 **Mara D. Neusel and Larry Smith,** Invariant theory of finite groups, 2002

93 **Nikolai K. Nikolski,** Operators, functions, and systems: An easy reading. Volume 2: Model operators and systems, 2002

92 **Nikolai K. Nikolski,** Operators, functions, and systems: An easy reading. Volume 1: Hardy, Hankel, and Toeplitz, 2002

91 **Richard Montgomery,** A tour of subriemannian geometries, their geodesics and applications, 2002

90 **Christian Gérard and Izabella Łaba,** Multiparticle quantum scattering in constant magnetic fields, 2002

89 **Michel Ledoux,** The concentration of measure phenomenon, 2001

88 **Edward Frenkel and David Ben-Zvi,** Vertex algebras and algebraic curves, second edition, 2004

TITLES IN THIS SERIES

87 **Bruno Poizat,** Stable groups, 2001

86 **Stanley N. Burris,** Number theoretic density and logical limit laws, 2001

85 **V. A. Kozlov, V. G. Maz′ya, and J. Rossmann,** Spectral problems associated with corner singularities of solutions to elliptic equations, 2001

84 **László Fuchs and Luigi Salce,** Modules over non-Noetherian domains, 2001

83 **Sigurdur Helgason,** Groups and geometric analysis: Integral geometry, invariant differential operators, and spherical functions, 2000

82 **Goro Shimura,** Arithmeticity in the theory of automorphic forms, 2000

81 **Michael E. Taylor,** Tools for PDE: Pseudodifferential operators, paradifferential operators, and layer potentials, 2000

80 **Lindsay N. Childs,** Taming wild extensions: Hopf algebras and local Galois module theory, 2000

79 **Joseph A. Cima and William T. Ross,** The backward shift on the Hardy space, 2000

78 **Boris A. Kupershmidt,** KP or mKP: Noncommutative mathematics of Lagrangian, Hamiltonian, and integrable systems, 2000

77 **Fumio Hiai and Dénes Petz,** The semicircle law, free random variables and entropy, 2000

76 **Frederick P. Gardiner and Nikola Lakic,** Quasiconformal Teichmüller theory, 2000

75 **Greg Hjorth,** Classification and orbit equivalence relations, 2000

74 **Daniel W. Stroock,** An introduction to the analysis of paths on a Riemannian manifold, 2000

73 **John Locker,** Spectral theory of non-self-adjoint two-point differential operators, 2000

72 **Gerald Teschl,** Jacobi operators and completely integrable nonlinear lattices, 1999

71 **Lajos Pukánszky,** Characters of connected Lie groups, 1999

70 **Carmen Chicone and Yuri Latushkin,** Evolution semigroups in dynamical systems and differential equations, 1999

69 **C. T. C. Wall (A. A. Ranicki, Editor),** Surgery on compact manifolds, second edition, 1999

68 **David A. Cox and Sheldon Katz,** Mirror symmetry and algebraic geometry, 1999

67 **A. Borel and N. Wallach,** Continuous cohomology, discrete subgroups, and representations of reductive groups, second edition, 2000

66 **Yu. Ilyashenko and Weigu Li,** Nonlocal bifurcations, 1999

65 **Carl Faith,** Rings and things and a fine array of twentieth century associative algebra, 1999

64 **Rene A. Carmona and Boris Rozovskii, Editors,** Stochastic partial differential equations: Six perspectives, 1999

63 **Mark Hovey,** Model categories, 1999

62 **Vladimir I. Bogachev,** Gaussian measures, 1998

61 **W. Norrie Everitt and Lawrence Markus,** Boundary value problems and symplectic algebra for ordinary differential and quasi-differential operators, 1999

60 **Iain Raeburn and Dana P. Williams,** Morita equivalence and continuous-trace $C^*$-algebras, 1998

59 **Paul Howard and Jean E. Rubin,** Consequences of the axiom of choice, 1998

58 **Pavel I. Etingof, Igor B. Frenkel, and Alexander A. Kirillov, Jr.,** Lectures on representation theory and Knizhnik-Zamolodchikov equations, 1998

For a complete list of titles in this series, visit the AMS Bookstore at **www.ams.org/bookstore/**.